Study Guide for
Ingraham and Ingraham's

INTRODUCTION TO MICROBIOLOGY

Second Edition

Jay M. Templin
Montgomery County Community College

Australia • Canada • Mexico • Singapore • Spain • United Kingdom • United States

For more information, contact:

BROOKS/COLE
511 Forest Lodge Road
Pacific Grove, CA 93950
USA
www.brookscole.com

Printed in the United States of America.

5 4 3 2 1

ISBN 0-534-55225-0

CONTENTS

PREFACE

This study guide contains a series of features to facilitate your learning of not only the facts but also the concepts and processes of microbiology. These features are:

Chapter Outline: The outline for the corresponding chapter of the textbook is found at the beginning of each chapter of the study guide. It lists the major topics that you will cover.

Key Terms: A list of the most important words from the text chapter immediately follows the chapter outline.

Study Tips: This is a new feature being introduced in this edition of the study guide. These are practical tips and suggestions on how to make your study habits and style more effective and efficient.

Study Exercises: Each study guide chapter presents a series of exercises. Each exercise contains questions that relate to a topic in the textbook. The topic is listed at the beginning of the exercise. This reference will help you to locate the relevant information in the textbook as you use the study guide.

Each exercise offers a series of questions that can be answered as you study the corresponding section in the textbook. The questions vary in format, ranging from short answer to matching to true-false. The questions provide an opportunity to actively practice as you study the content in the textbook. They are designed to promote your learning of microbiology. In this edition, a few discussion questions have also been included to sharpen your analytical and critical thinking skills in studying microbiology concepts.

Multiple Choice: The study exercises are followed by multiple choice questions that provide a general review of the chapter's content.

Answer Key: The chapter concludes with the correct answers for each of the exercises in the study guide. After answering the questions, you can check this key to learn about your progress toward mastery of the content in the textbook.

Jay M. Templin
Montgomery County Community College
Pottstown, PA

Chapter 1
The Science of Microbiology

Key Terms

microbiology
pathogen
eukaryotic
prokaryotic
bacteria
archaea
algae
fungi
protozoa
helminth
spontaneous generation
scientific method
immunity

Chapter 1

vaccination
pasteurization
chemotherapy
drug
antibiotic
genetic engineering

Study Tips

1. To start studying Chapter 1, scan the chapter for major topics and subtopics. Also, read the chapter summary before studying the chapter in depth. This approach can provide a quick overview of the chapter's content.

2. Learning vocabulary is a major challenge for any student in a science course. After studying Chapter 1 write your own brief definition for each key term in the chapter.

3. Answer the review questions at the end of the chapter. Your instructor can supply you with the correct answers once you have tried to answer the questions on your own.

4. Take some study time to look over all of the chapters in your textbook. Can you find where each topic noted in Chapter 1 is covered in more detail? Where is the topic of recombinant DNA technology covered again? Are there other discussions related to medical microbiology? Can you find a more in-depth description about the differences between eukaryotic and prokaryotic cells?

5. Have you taken other biology courses before microbiology? Can you relate some of the concepts of these courses to the content you are studying now? For example did you learn about the immune system when studying human anatomy and physiology? Where is insulin normally produced in the body? Where does *E. coli* normally inhabit the human body?

The Unseen World and Our World

A. Label each one of the following statements as true or false. If false, correct the statement.

   1. The bubonic plague was caused by a bacterium. T

   2. Infected rats and fleas carried the microorganism that caused the bubonic plague in Europe during the Middle Ages. T

   3. Today, throughout the world, the bubonic plague does not occur as it did during the Middle Ages. F

   4. The potato blight in Ireland in the 1800s was caused by a bacterium. F – fungus

5. In the 1500s, the Incas of South America had not developed immunity to smallpox and measles when first exposed to the microorganism causing it.

6. Tetanus and gas gangrene are viral diseases.

7. A pathogen is a microorganism that has a beneficial effect on humans.

B. Complete each of the following statements with the correct term or terms.

1. The study of how microorganisms affect the earth and its atmosphere is called ________ microbiology.

2. Vaccines have been developed to protect humans against ________, microorganisms that can cause disease.

3. AIDS is an abbreviation for ________.

4. Using microorganisms to produce a food supply is a goal of ________ microbiology.

5. The first human use of microbes was to make and preserve ________.

6. Microorganisms that are used to kill insects are natural ________.

7. Water is made safe for human consumption through the advances of ________ microbiology.

The Scope of Microbiology

A. Complete each of the following statements with the correct term.

1. Microorganisms are currently divided into ________ subgroups.

2. The ________ are the recently recognized subgroup of microorganisms.

3. A primary distinction among the subgroups of microorganisms is in their ________ structure.

4. Viruses are ________, meaning that they lack a cell structure.

5. Some worm species, the ________, are traditionally a part of the study of microbiology.

B. Match each description to the correct group of organisms.

1. phytoplankton are members	A. archaea
2. most are unicellular, procaryotic	B. algae

3. worms of the animal kingdom
4. nucleic acid core, protein coat
5. one member causes malaria
6. one type causes AIDS
7. most form a mycelium
8. spheres and rods are characteristic shapes
9. most complex unicellular structure
10. one kind causes syphilis
11. kelp belongs to this group
12. discovered in the 1970s

C. bacteria
D. fungi
E. helminths
F. protozoans
G. viruses

A Brief History of Microbiology

A. Complete each of the following statements with the correct term or terms.

1. The most powerful magnification of Leeuwenhoek's microscopes was ________X.
2. Leeuwenhoek referred to the microorganisms he studied as ________.
3. Hooke experimented with ________ microscopes.
4. The most powerful range of Hooke's microscopes was from _______ to _______X.
5. Redi conducted experiments testing the theory of ________.
6. Pasteur tested the theory of spontaneous generation by using flasks with ________ necks.
7. ________ are structures that some bacteria form, making them resistant to heat.
8. A ________ culture contains only one kind of microorganism.
9. Koch developed four ________ to prove that a particular microorganism causes a disease.
10. ________ ________ is the bacterium causing anthrax.
11. A ________ contains a weakened form of a microorganism that can cause a disease.

12. Jenner is most noted for studying the disease ________.

13. The term "vacca" is a Latin term meaning ________.

14. Jenner's work lead to extended principles of ________ toward other diseases.

15. By the process ________ most microorganisms are killed by brief exposure to heat.

Microbiology Today

A. Label each one of the following statements as true or false. If false, correct it.

1. Pasteur articulated the principle of selective toxicity.

2. Salvarsan was the first drug used to successfully treat syphilis.

3. Erythromycin was the first medically useful antibiotic.

4. Iwanowsky observed that bacteria could pass through the pores in the filters of his experiment.

5. The tobacco mosaic virus infects plants.

6. Bacteria are easy to culture.

7. Bacteria multiply rapidly.

8. *Staphylococcus aureus* is the microorganism used in DNA recombinant studies.

9. Under proper conditions bacteria can double their numbers every 20 minutes.

10. Recombinant DNA experiments led to the technology to produce penicillin.

Discussion Questions

1. What are some possible dangers to human populations when new strains of microorganisms are produced through genetic engineering?

2. How do other courses in science (i.e., chemistry) relate to what you are currently studying?

3. Did spontaneous generation occur once on the earth in the past?

Chapter 1

Multiple Choice: Review

1. Lister used ________ as an antiseptic.

    A. alcohol
    B. carbolic acid
    C. heat
    D. penicillin

2. *Yersinia pestis* is the causative agent of the disease

    A. bubonic plague.
    B. pneumonia.
    C. syphilis.
    D. typhoid fever.

3. Which one of the following subgroups has a eukaryotic, complex cell structure?

    A. archaea
    B. bacteria
    C. protozoa
    D. viruses

4. Select the incorrect statement about viruses.

    A. They are living outside a host.
    B. They contain a nucleic acid as part of their structure.
    C. They contain a protein as part of their structure.
    D. They lack a cellular structure.

5. Select the incorrect statement about eukaryotic cell structure.

    A. A true nucleus is lacking in the cells.
    B. Algae have this kind of structure.
    C. Bacteria lack this kind of structure.
    D. Many organelles are present.

6. Select the incorrect characteristic about algae.

    A. eukaryotic
    B. photosynthetic
    C. some are unicellular
    D. some are fungi

7. Select the incorrect statement about protozoans.

D

A. Cilia are used by some for movement.
B. Most are unicellular
C. Their cells have numerous organelles.
D. They are similar viruses.

8. Select the incorrect association.

D

A. Ehrlich-contributed to the development of chemotherapy
B. Iwanowski-studied viruses
C. Koch-developed postulates
D. Pasteur-proved the theory of spontaneous generation

9. *Bacillus anthracis* causes

B

A. a disease in algae.
B. a disease in cattle.
C. smallpox.
D. syphilis.

10. The first drug named the wonder drug was

B

A. erythromycin.
B. penicillin.
C. tetracycline.
D. sulfa drug A.

Answers

The Unseen World and Our World

A. 1 True
2. True
3. False; The bubonic plague still occurs in the western and southwestern United States.
4. False; The potato blight was caused by a fungus.
5. True
6. False; Tetanus and gas gangrene are caused by bacteria.
7. False; A pathogen infects the human body and causes a disease.

B. 1. environmental 2. pathogens 3. acquired immuno-deficiency syndrome 4. agricultural 5. food 6. pesticides 7. environmental

Chapter 1

The Scope of Microbiology

A. 1. six 2. achaea 3. cell 4. acellular 5. Helminths

B. 1. B 2. C 3. E 4. G 5. F 6. G 7. D 8. C 9. F 10. C 11. B 12. A

A Brief History of Microbiology

A. 1. 266 2. animalcules 3. two-lensed 4. 300 to 500 5. spontaneous generation 6. curved 7. endospores 8. pure 9. postulates 10. *Bacilllus anthracis* 11. vaccine 12. smallpox 13. cow 14. immunization 15. pasteurization

Microbiology Today

A. 1. False: Ehrlich articulated the principle of selective toxicity.
2. True
3. False; Penicillin was the first medically useful antibiotic.
4. False; Iwanowsky observed that viruses could pass through the pores in his filters.
5. True
6. True
7. True
8. False; *Escherichia coli* is the microorganism used in DNA recombinant studies.
9. True
10. False; Penicillin was discovered before the advent of DNA recombinant technology.

Discussion Questions

1. Start with the idea that genetic engineering could produce a mutant strain of microorganism that could defeat the human immune system.

2. Are there applications from chemistry (i.e., molecules) or from human anatomy and physiology (i.e., immune system)?

3. Start with the evidence that life began with microorganisms over 3 billion years ago.

Multiple Choice: Review

1. B 2. A 3. C 4. A 5. A 6. D 7. D 8. D 9. B 10.B

Chapter 2
Basic Chemistry

- The Basic Building Blocks
  - Atoms
  - Elements
  - Molecules
    - Molecular Weight
    - Moles
- Chemical Bonds and Reactions
  - Covalent Bonds
  - Ionic Bonds
  - Hydrogen Bonds
  - Chemical Bonds
    - Reaction Rates
    - Enzymes
- Water
  - Special Properties of Water
  - Water as a Solvent
    - Colloids
    - Hydrophobic and Hydrophilic Interactions
    - Hydrogen and Hydroxide Ions
    - Acids, Bases, and Salts
    - The pH Scale
    - Buffers
- Organic Molecules
  - Carbon
  - Functional Groups
- Macromolecules
  - Proteins
    - Amino Acids
    - Peptide Bonds
    - Protein Structure
    - Denaturation
  - Nucleic Acids
    - DNA
    - RNA
  - Polysaccharides
    - Sugars
    - Glucose-containing Polysaccharides
  - Lipids
    - Lipids Containing Fatty Acids
    - Lipids Not Containing Fatty Acids
- Summary

Chapter 2

Key Terms

atom
element
neutron
proton
electron
atomic number
atomic weight
energy shell
molecule
molecular formula
molecular weight
mole
Avogadro's number
chemical bond
covalent bond
ionic bond
cation
anion
hydrogen bond
chemical reaction
reactant
product
free energy
activation energy
enzyme
ribozyme
catalytic site
specific heat
solution
solute
solvent
hydrophobic
hydrophilic
dissociate
spheres of hydration
colloid
turbid
acid
base
salt
pH scale
buffer
organic molecule
hydrocarbon

functional group
macromolecule
polymer
monomer
dehydration synthesis
hydrolysis
protein
amino acid
peptide bond
primary structure
alpha helix
beta sheet
denaturation
sterilization
nucleic acid
nucleotide
purine
pyrimidine
polysaccharide
monosaccharide
disaccharide
lipid
fatty acid
phospholipid
sterol

Study Tips

1. Answer the several kinds of questions at the end of this chapter. Your instructor has the answers to these questions. You can check your mastery of the facts and concepts in this chapter after trying these questions and checking your answers.

2. Good students also learn to write additional questions, anticipating the kinds of questions that may be posed on exams. Trying writing your own questions from each section in the chapter. Try working with another student in your class and testing each other with your questions.

3. Many colleges supply stickball models and kits to simulate the molecules you are studying in this chapter. Ask your instructor about their availability at your school and use them as another tool to visualize the bonding patterns discussed in Chapter 2.

4. Look through the upcoming chapters in the textbook. Can you find examples where the facts and principles of chemistry will apply to other topics of microbiology that you will study? Examples include metabolism, nutrition, and genetics.

5. Do you see applications of chemistry to the lab component of your course? What kinds of molecules do bacteria use as a source of energy in their cells? What is meant by the fermentation of carbohydrates?

The Basic Building Blocks

A. Complete each of the following statements with the correct term or terms.

1. The ________ is the subatomic particle found in energy shells outside the nucleus of the atom.

2. The __________ is the subatomic particle with a positive charge.

3. The _________ is the subatomic particle with no charge.

4. The innermost shell of an atom can hold __(number)__ or __(number)__ electrons.

5. An intact atom has the same number of protons and _________.

6. The atomic ________ is the number of protons and neutrons.

7. Oxygen has an atomic number of _________.

8. Oxygen usually has an atomic weight of _________.

9. The most common isotope of carbon has __________ protons and __________ neutrons.

10. The atomic number of chlorine is _________.

11. An element has 11 protons and 12 neutrons in its nucleus. Its atomic weight is _________.

12. An element has 8 protons and 10 neutrons in its nucleus. Its atomic number is _________.

13. An element has an atomic number of 8. The number of electrons in its outermost energy shell is _________.

14. An element has an atomic number of 7. The number of electrons in its innermost energy shell is _________.

15. The molecular weight of water is _________.

B. Label each of the following statements as true or false. If false, correct it.

1. An intact atom is electrically neutral.

2. Electrons are found in the nucleus of an atom.

3. The electron is not found in the nucleus of an atom.

4. The neutron is found in the nucleus of an atom.

5. The electrons in the innermost energy shell of an atom are valence electrons.

6. If an atom has an atomic number of 6, it has 3 valence electrons.

7. The atomic weight of carbon is 6.

8. A mole of water weighs 24 grams.

9. A mole of hydrogen weighs 5 grams.

10. Sulfur is a trace element.

C. Refer to Figure I to answer the following questions.

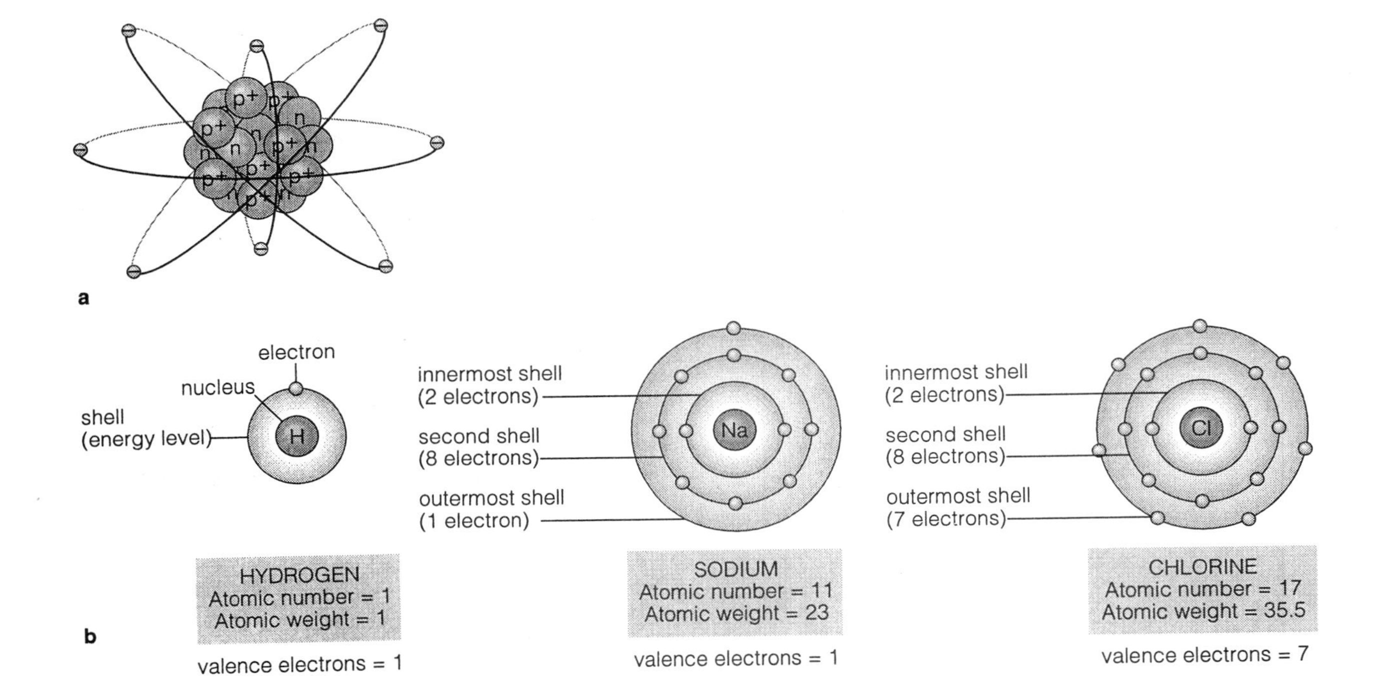

Chapter 2

1. Atom X has an atomic number of ________.
2. Atom X represents the element ________.
3. Atom Y has ________ protons in its nucleus.
4. Atom Y represents the element ________.
5. Atom Z has __________ protons in its nucleus.
6. If atom Z has 18 neutrons, its atomic weight is ________.
7. How many more electrons does atom Z need for a full outer shell configuration?
8. Atom Z represents the element _________.

Chemical Bonds and Reactions

A. Label each of the following as describing a covalent bond, ionic bond, or hydrogen bond.

1. If forms between water molecules.
2. It is formed by the activity of an electron donor.
3. It is formed by the sharing of electrons.
4. This bond exists in sodium chloride.
5. It is formed by the activity of an electron acceptor.
6. It can be single, double, or triple.
7. This bond is found in polar and nonpolar molecules.
8. This bond is formed by the attraction between oppositely charged particles.
9. An oxygen atom can form two of this type of bond.
10. This bond stabilizes a large molecule such as DNA, helping to stabilize its shape.
11. This is the most stable bond.
12. This bond involves positive sodium particles.
13. It is an extremely weak bond.

14. NaCl has this kind of bond.

15. Carbon forms four of these bonds.

B. Complete each of the following statements with the correct term or terms.

1. A chemical reaction changes reactants into ________.

2. Reactions only occur between substances with enough energy to enter an _________ state.

3. _________ are substances that increase the rate of a reaction in living systems.

4. _________ are enzyme RNA molecules.

5. Enzyme names end in the suffix ________.

6. *E. coli* has about _________ different enzymes.

Water

A. Label each of the following statements as true or false. If false, correct it.

1. About 40 percent of the weight of a bacterial cell consists of water.

2. Hydrogen bonds increase the heat of vaporization of water.

3. Water is a substance with a high specific heat.

4. As a solid, ice is more dense than liquid water.

5. Compounds that do not interact with water are hydrophilic.

6. Compounds that dissolve in water are hydrophobic.

7. Hydrogen ions have a positive charge.

8. The solution with a hydrogen ion concentration of 0.001 grams per liter has a pH of 4.

9. Acids are proton acceptors in a solution.

10. Buffers are substances that tend to change the pH of a solution.

B. Select the correct choice for each statement.

1. Hydrogen bonds make water

   A. adhesive
   B. cohesive

2. NaCl is the _________ of a saltwater solution.

   A. solute
   B. solvent

3. NaCl is

   A. hydrophilic
   B. hydrophobic

4. Colloid suspensions are usually

   A. clear
   B. cloudy

5. Water-fearing substances are

   A. hydrophilic
   B. hydrophobic

6. The charge of a hydroxide ion is

   A. negative
   B. positive

7. Acid solutions contain more ________ ions.

   A. hydrogen
   B. hydroxide

8. A basic solution has a pH _________ than 7.

   A. higher
   B. lower

9. The pH of 4 is _________ times more acidic that a pH of 6.

   A. 10
   B. 100

10. The pH of 6 is __________ times more basic than a pH of 3.

    A. 100
    B. 1000

11. Hydrophilic means

    A. water fearing
    B. water loving

12. NaCl is

    A. water fearing
    B. water loving

13. The more acidic pH is

    A. 5.4
    B. 5.6

14. The more basic pH is

    A. 8.4
    B. 8.8

15. Some bacteria have an intracellular pH as high as

    A. 7.2
    B. 8.5

Organic Molecules

A. Complete each of the following statements with the correct term.

1. Organic compounds contain at least the elements ________ and ________.
2. The functional group -COOH is called the ________ group.
3. The amino group contains the elements ________ and ________.
4. Molecules that contain the amino groups are called ________, as they readily accept protons.
5. A polar functional group makes a compound more _________ in water.

Chapter 2

Macromolecules

A. Complete the following statements about proteins with the correct term or terms.

1. The monomers of proteins are ________ ________.
2. The polymers of proteins are broken into monomers through ________, the addition of water at their chemical bonds.
3. Amino acids are linked by _________ bonds.
4. A _________ is two amino acids joined together.
5. Th L and D forms are different ________ of an amino acid.
6. ________ is the destruction of a protein's three-dimensional structure.
7. The ________ structure of a protein is its order of bonded amino acids.
8. Disulfide bonds establish a protein's ________ structure.
9. The ________ structure of a protein refers to how the different chains of molecules fit together.
10. The alpha helix and pleated sheet refer to a protein's ________ structure.

Refer to Figure II to answer questions 11 through 14.

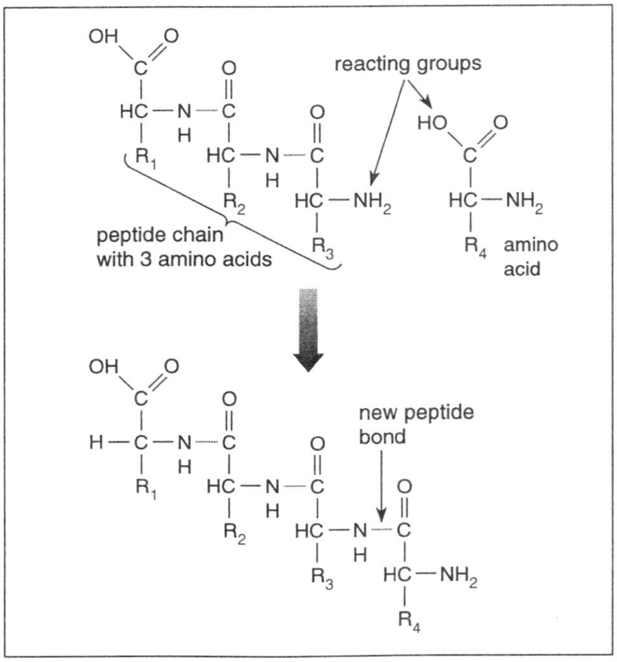

reacting groups
peptide chain with 3 amino acids
amino acid
new peptide bond

11. Name all of the different elements found in the amino acid.

12. Each peptide bond is formed with the loss of a molecule of ________.

13. The amino acids are joined to form a _________ chain.

14. What does the R group represent on each amino acid?

B. Complete each of the following statements about nucleic acids with the correct term.

1. Part of a ribosome's composition consists of the nucleic acid ________.

2. _________ is the nucleic acid that encodes the genetic information in a cell.

3. _________ are the monomers of nucleic acids.

4. Cytosine and thymine are _________, single-ringed bases.

5. Adenine and guanine are _________, double-ringed bases.

6. __________ is the major carrier molecule of cellular energy.

7. Nucleotides are joined in a single-stranded nucleic acid by _________ bonds.

8. DNA has the shape of a _________ helix.

9. In DNA the base adenine always bonds to _________.

10. In DNA the base cytosine always bonds to _________.

11. In DNA the base thymine always bond to _________.

12. In DNA the base guanine always bonds to _________.

C. Match each of the following descriptions to the correct kind of carbohydrate.

1. glucose is one example
2. cellulose is one example
3. reserve energy storage
4. reserve energy storage molecule in animals
5. two monosaccharides bonded

A. monosaccharide
B. disaccharide
C. polysaccharide molecule in plants

6. lactose is an example

7. starch is an example

8. galactose is an example

9. simplest sugar

10. ring form can be alpha or beta

11. the largest carbohydrate

12. simple sugar

D. Refer to Figure III to answer questions 1 through 6.

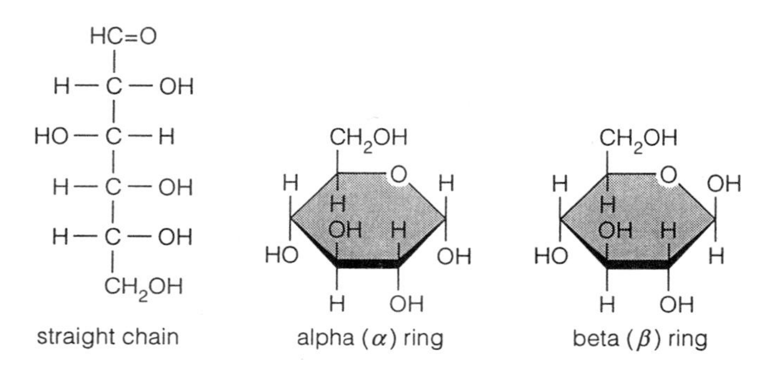

1. How many carbon atoms are in a molecule of the straight chain?

2. How many hydrogen atoms are in a molecule of the straight chain?

3. How many oxygen atoms are in a molecule of the straight chain?

4. The straight chain represents a molecule of ________.

5. The alpha and beta rings each represent a molecule of ________.

6. If the two ring forms bond, there is a loss of a molecule of ________ between them.

E. Label each of the following statements about lipids as true or false. If false, correct it.

1. The building blocks of lipids are chemically broken down by a dehydration synthesis.

2. Lipids are polar molecules.

3. Fatty acids can be saturated or unsaturated.

4. A phospholipid molecule contains three fatty acids.

5. Fatty acids are joined to glycerol in the lipid molecule by a ester linkage.

6. The phosphate group forms the nonpolar head of a phospholipid.

7. Sterols contain fatty acids.

8. Cholesterol is a sterol.

9. Phospholipid molecules tend to form membranes.

10. Lipids do no make up the genetic systems of cells.

11. Fatty acids with double bonds are unsaturated.

12. Most bacteria have sterols.

Discussion Questions:

1. Life consists of complex molecules. However, these molecules consist of the simplest elements. How did life evolve from this pattern?

2. How are the principles of this chapter important to the topic of bacterial nutrition?

Multiple Choice: Review

1. An element has an atomic number of 13 and an atomic weight of 27. The number of protons in its nucleus is

   A. 13
   B. 14
   C. 27
   D. 28

2. An atom has an atomic number of 13. The number of electrons in its outermost energy shell is

   A. 2
   B. 3
   C. 5
   D. 7

3. Select the best description for the water molecule.

   A. sharing of electrons, polar
   B. sharing of electrons, nonpolar
   C. transfer of electrons, polar
   D. transfer of electrons, nonpolar

4. Select the incorrect characteristic about water.

   A. Each of molecule consists of three atoms.
   B. It is a solvent.
   C. It has a boiling point of 212°C.
   D. It is found in the cells of microorganisms.

5. Base molecules dissociate to form ________ ions.

   A. hydrogen
   B. hydroxide

6. A pH of 4 is _________ times more acidic than a pH of 6.

   A. 10
   B. 100
   C. 1000
   D. 10000

7. Select the correct association.

   A. amino acids/polysaccharides
   B. glucose/nucleotide
   C. monosaccharide/glucose
   D. RNA/lipid

8. Select the element found in proteins that is not found in glucose.

   A. carbon
   B. hydrogen
   C. nitrogen
   D. oxygen

9. Select the correct statement about DNA.

   A. It contains the sugar ribose.
   B. It is usually single-stranded.
   C. It makes RNA.
   D. Some of its nucleotide contain uracil.

10. The monosaccharides of sucrose are

    A. glucose and glucose.
    B. glucose and fructose.
    C. lactose and galactose.
    D. lactose and fructose.

Answers

The Basic Building Blocks

A. 1. electron 2. proton 3. neutron 4. one or two 5. electrons 6. weight 7. eight 8. sixteen 9. six, six 10. seventeen 11. twenty-three 12. eight 13. six 14. two 15. eighteen

B. 1. True
   2. False; Electrons are in the shells around the nucleus.
   3. True
   4. True
   5. False; The electrons in the outermost shell of the atom are the valence electrons.
   6. False: It has four valence electrons.
   7. False; The atomic number of carbon is six.
   8. False; A mole of water weighs eighteen grams.
   9. False; A mole of hydrogen weighs two grams.
   10. True

C. 1. one 2. hydrogen 3. eleven 4. sodium 5. seventeen 6. thirty-five 7. one 8. chlorine

Chemical Bonds and Reactions

A. 1. hydrogen 2. ionic 3. covalent 4. ionic 5. ionic 6. covalent 7. covalent 8. ionic 9. covalent 10. hydrogen 11. covalent 12. ionic 13. hydrogen 14. ionic 15. covalent

B. 1. products 2. activation 3. enzymes 4. ribozymes 5. ase 6. one thousand

Chapter 2

Water

A.
1. False; Water is about 70 percent of the weight of bacterial cells.
2. True
3. True
4. False; Ice is less dense than liquid water.
5. False; Compounds that do not interact with water are hydrophobic.
6. False; Compounds that dissolve in water are hydrophilic.
7. True
8. False; This solution has a pH of 3.
9. False; Acids are proton donors in solution.
10. False; Buffers tend to stabilize the pH of a solution.

B. 1. B 2. A 3. A 4. B 5. B 6. A 7. A 8. A 9. B 10. B 11. B 12. B 13. A 14. B 15. B

Organic Molecules

A. 1. carbon and hydrogen 2. carboxyl 3. nitrogen and hydrogen 4. bases 5. soluble

Macromolecules

A. 1. amino acids 2. hydrolysis 3. peptide 4. dipeptide 5. isomers 6. denaturation 7. primary 8. tertiary 9. quaternary 10. secondary 11. carbon, hydrogen, oxygen, nitrogen 12. water 13. polypeptide 14. The R group is the variable part of the amino acid.

B. 1. RNA 2. DNA 3. nucleotides 4. pyrmidines 5. purines 6. ATP 7. phosphodiester 8. double 9. thymine 10. guanine 11. adenine 12. cytosine

C. 1. A 2. C 3. C 4. C 5. B 6. B 7. C 8. A 9. A 10. A 11. C 12. A

D. 1. six 2. twelve 3. six 4. glucose 5. glucose 6. water

E.
1. False; They are broken down by hydrolysis.
2. False; Lipids are nonpolar molecules and do not dissolve in water.
3. True
4. False; A phospholipid molecule contains two fatty acids.
5. True
6. False; The phosphate group forms the polar head of the molecule.
7. False; Sterols lack fatty acids.
8. True
9. True
10. True
11. True
12. False; Most bacteria lack sterols

Discussion Questions

1. Atoms can be assembled into more complex subunits. It is the arrangement and repeating pattern of these subunits that led to the formation of the complex macromolecules of life.

2. Study protein synthesis, starch hydrolysis, and the fermentation of various sugars. You will study these in more depth in the future. However, look ahead in the text now and start to become familiar with their broad patterns.

Multiple Choice: Review

1. A 2. B 3. A 4. C 5. B 6. B 7. D 8. C 9. C 10. B

Chapter 3
Methods of Studying Microorganisms

Viewing Microorganisms
- Properties of Light
- Microscopy
- The Compound Light Microscope
- Wet Mounts
- Stains
- Light Microscopy: Other Ways to Achieve Contrast
- Electron Microscopy
- Uses of Microscopy

Cultivating Microorganisms
- Obtaining a Pure Culture
- Sterilization
- Isolation
- Growing a Pure Culture
- Providing a Suitable Environment

Preserving a Pure Culture
Microorganisms that Cannot be Cultivated in the Laboratory
Summary

Key Terms

electromagnetic waves
gamma rays
light
reflection
transmission
absorption
diffraction
refraction
magnification
resolution
numerical aperture
compound light microscope
wet mount
vital stain
heat fixation
simple stain
differential stain
special stain
electron microscope
freeze-fracturing

freeze-etching
pure culture
filtration
incubation
fastidious
sterilization
autoclave
serial dilution
complex medium
differential medium
enrichment culture
stock culture

Study Tips

1. Continue to scan each chapter for major topics and subtopics before studying the chapter content in depth.

2. After studying the chapter, list the key terms and write a definition for each term in your own words.

3. Outlining the chapter in your own words is another means of active practice to learn the content of the chapter. Start to make a list of topics and subtopics. Under each topic and/or subtopic list several key ideas that you have learned.

4. Answer the review questions and correlation questions at the end of the chapter. Your instructor can supply you with the correct answers once you have tried to answer the questions on your own.

5. Use your experiences in lab to help you understand the concepts you are learning in lecture. Use of the microscope will help you to understand the function of its parts as well as the concepts of magnification and resolution.

Viewing Microorganisms

A. Label each of the following statements as true or false. If false, correct it.

   1. Gamma rays are a visible part of the electromagnetic spectrum.

   2. Blue light has longer wavelengths than red light.

   3. Yellow and green are intermediate wavelengths of light.

   4. A solution of red dye absorbs the red component of white light.

   5. At room temperature pure water has a refractive index of 1.0.

6. Magnification by a lens is increased by making the lens more concave.

7. When light is transmitted it bounces off the surface of an object.

8. Diffraction is the straightening out of light rays.

9. As light enters a denser medium, it slows down.

10. The refractive index of water is greater than the refractive index of air.

B. Short Answer

1. Name the two ways that magnification can be increased in a compound light microscope.

2. How is the contrast of a microscopic image created?

3. ________ is the distinguishing of two points of an image.

4. List the three factors that increase the resolving power of a microscope.

5. How does the wavelength of light affect the resolving power by a microscope?

6. How does a beam of electrons in microscopy affect resolution?

7. What are the lenses used in a compound light microscope?

8. What is the function of the condenser of the compound light microscope?

9. How does changing the objective lens change the total magnification of the compound light microscope?

10. What is the advantage of low power over high power when viewing specimens with the microscope?

C. Matching

1. acid-fast stain
2. Leifson stain
3. Gram stain
4. hanging drop mount
5. negative stain

a. simplest way to prepare a specimen
b. iodine used as a mordant
c. sulfuric acid used to decolorize
d. used to reveal bacterial capsules
e. used to reveal flagella

6. wet mount — f. uses a seal of petroleum jelly

D. Short Answer

1. List the four steps of the Gram stain, in their correct order.
2. List the four steps of the acid-fast stain, in their correct order.
3. List the five steps of the flagellar stain, in their correct order.

E. Match each of the following descriptions to the correct kind of microscopy.

1. generated by the interference of light rays	A. TEM
2. used to reveal antibodies	B. SEM
3. uses a pair of prisms	C. Nomarsky
4. uses freeze-fracturing	D. fluorescence
5. can magnify only up to 10,000X	E. darkfield

Cultivating Microorganisms

A. Label each of the following statements as true or false. If false, correct it.

1. A mixed culture contains many different kinds of microorganisms.
2. Sterilization is the elimination of all microorganisms.
3. Moist heat is more effective at killing microorganisms than dry heat.
4. At 15 pounds per square inch water boils at 100 degrees Celsius.
5. Glass and metal instruments can be sterilized by dry heat.
6. Filtration usually remove viruses from liquids.
7. By filtration the filters to remove microbial cells usually have pores of 0. 8 micrometers.
8. Sodium chloride is bleach.
9. The serial dilution is used to increase the concentration of microbial cells in a sample.
10. The streak-plate method is used in a test tube of nutrient broth.

11. A suspension with 1200 cells per milliliter is diluted by 3 steps of a serial dilution. Each step dilutes the cell number by one-tenth. After the three steps the cell concentration is 1.2 cells per milliliter.

12. From question #11, two steps from the original suspension produce a cell concentration of 12 cells per milliliter.

B. Match each description to the correct kind of growth medium.

1. used to culture fastidious microorganisms
2. isolates endospore-forming bacteria from a large, complex population
3. MacConkey agar is one example
4. blood agar is one example
5. favors the growth of some microorganisms and suppresses others
6. medium with only a pH indicator is one example
7. nutrient agar is one example
8. nutrient broth is one example
9. *E. coli* grows in a simple kind of medium.
10. used to isolate *Salmonella typhi*
11. used for *S. pyrogenes*
12. exact chemical composition known

a. complex
b. defined
c. differential
d. enrichment
e. selective
f. selective-differential

C. Complete each of the following statements with the correct term or terms.

1. Most bacteria grow over a temperature range of ________ degrees Celsius.
2. *Escherichia coli* lives in the human ________.
3. *E. coli* grows best at a temperature of ________ degrees Celsius.
4. Fungi grow best at a pH range of ________ to ________.
5. Facultative anaerobes can grow in the ________ or ________ of oxygen.

6. Anaerobes cannot grow in the presence of _______.

7. ________ cultures are maintained for study and reference.

8. ________ is freeze-drying.

9. ________ is the removal of all water when frozen.

10. *Treponema pallidum* is the causative agent of ________.

11. ________ *leprae* causes leprosy.

12. Viruses are grown in ________ cultures.

Discussion Questions

1. How does your background in physics and chemistry help you to understand the concepts in this chapter?

2. You will be studying more pathogens in the upcoming chapters. List some of the ones you read about in Chapter 3. Name the disease caused by each one.

3. What are some of the advantages of using microorganisms for study in the laboratory?

Multiple Choice: Review

1. Select the color with the longest wavelengths from the visible part of the electromagnetic spectrum.

   A. blue
   B. green
   C. red
   D. yellow

2. If all of the light is transferred to an object when it strikes it, the light is

   A. absorbed.
   B. reflected.
   C. translated.
   D. transmitted.

3. The resolving power of a microscope is determined in part of the refractive index of material

   A. between the objective lens and specimen.
   B. between the ocular and objective lens.
   C. in the lenses.
   D. in the specimen.

4. The primary stain for the Gram stain is

   A. acetone.
   B. alcohol.
   C. gentian violet.
   D. safranin.

5. The decolorizing agent for the acid-fast stain is

   A. carbolfuchsin.
   B. heat.
   C. methylene blue.
   D. sulfuric acid.

6. Which kind of microscopy uses two prisms?

   A. darkfield
   B. fluorescence
   C. Nomarsky
   D. phase-contrast

7. The requirements for sterilizing most microorganisms in the autoclave is

   A. 100°C, 15 minutes
   B. 100°C, 20 minutes
   C. 121°C, 15 minutes
   D. 121°C, 20 minutes

8. A sample of broth culture of bacteria is diluted 1000 times. One ml of this diluted culture forms 19 colonies on a nutrient agar surface. The original concentration of cells per ml was

   A. 1900
   B. 19000
   C. 190000
   D. 1900000

9. SPS agar is a ________ medium.

   A. complex
   B. hazardous
   C. differential
   D. selective

10. The purpose of lyophilization is to

   A. add water to an environment.
   B. hold the temperature constant.
   C. remove all moisture.
   D. stabilize the pH.

Answers

Viewing Microorganisms

A.
1. False; Gamma rays have the shortest wavelengths and are invisible.
2. False; Blue light has the shortest, visible wavelengths.
3. True
4. False; It reflects the red wavelengths and this is what you see.
5. False; At room temperature pure water has a refractive index of 1.33.
6. False; Magnification is increased by making the lens more convex.
7. False; When light is transmitted it passes through the surface of an object.
8. False; Diffraction is the bending of light rays.
9. True
10. True

B.
1. Magnification is increased by making the lens more convex or by bringing the object closer to the lens.
2. Some parts of a viewed image absorb light more than others. These differences produce the contrast.
3. resolution
4. Resolution increases by increasing the size of the lenses, increasing the refractive index, and using immersion oil.
5. Smaller wavelengths increase the resolving power.
6. An electron beam increases resolution.
7. The lenses used in the compound light microscope are the ocular lens and several objective lenses.
8. The condenser concentrates and directs light through the stage mounting the specimen.
9. The ocular lens is constant, usually 10x. The objective lens used can be several choices: 4x, 10x, or 40x. Total magnification is ocular times objective: i.e., 10x times 40x.
10. Use of a low power does not provide maximum magnification but does allow scanning of most or all of the specimen being viewed.

C. 1. C 2. E 3. B 4. F 5. D 6. A

D. 1. The specimen is stained with a primary stain such as gentian violet. Iodine is applied as a mordant. A decolorizing agent is added. The specimen is counterstained with safranin.

2. The specimen is stained with the primary stain carbolfuchsin. The microslide is heat-dried. The decolorizing solution of sulfuric acid in ethanol is added. The specimen is counterstained with methylene blue.

3. A bacterial suspension is fixed with formalin. The slide is air-dried without heating. A mixture of tannic acid and rosaniline dyes is added to the slide. The excess stain is washed off the slide by flooding with water. The slide is air-dried.

E. 1. E 2. D 3. C 4. A 5. B

Cultivating Microorganisms

A. 1. True
2. True
3. True
4. False; At 15 pound per square inch water boils at 121 degrees Celsius.
5. True
6. False; Viruses pass through the pores of the filter.
7. False; The pores are about 0. 45 micrometers.
8. False; Sodium hypochlorite is bleach.
9. False; A serial dilution reduces the cell population to a countable number when plated on a nutrient agar surface.
10. False: The streak plate method is done on a nutrient agar surface in a Petri dish.
11. True
12. True

B. 1. b 2. d 3. f 4. c 5. e 6. c 7. a 8. a 9. b 10. e 11. c 12. b

C. 1. 40 2. intestine 3. 374. 4. 5 to 6. 0 5. presence or absence 6. oxygen 7. stock 8. lyophilization 9. sublimation 10. syphilis 11. *Mycobacterium* 12. tissue

Discussion Questions

1. How does the physics of light help you to understand microscopy? Explain what you know about the wavelength of light. How does chemistry relate to the staining of bacteria or the methods of chemical control? Explain what you know about alcohol, acetone, etc.

2. Start with the species of *Mycobacterium* and *Salmonella*.

3. Concentration on the ease to cultivate them in the lab. What does it mean if their haploid?

Multiple Choice: Review

1. C 2. A 3. A 4. C 5. D 6. C 7. C 8. B 9. D 10. C

Chapter 4
Prokaryotic and Eukaryotic Cells

- The Prokaryotic Cell
  - Structure of the Bacterial Cell
    - The Capsule and Envelope
      - Capsule
      - Outer Membrane
      - Periplasm
      - The Cell Wall
        - Structure
        - Cell Shape
        - Turgor Pressure
        - Mycoplasmas
        - L Forms
- The Cytoplasmic Membrane
    - Appendages
      - Pili
      - Flagella
      - Axial Filaments
    - Cytoplasm
      - Nucleoid
      - Ribosomes
      - Storage Granules
      - Other Inclusions
    - Endospores
  - Structure of the Archaeon Cell
- The Eukaryotic Cell
  - Structure
    - Appendages
    - Cell Wall
    - Cytoplasmic Membrane
    - Cytoplasm and Its Contents
      - Cytoskeleton
      - Nucleus
      - Cytomembrane System and Ribosomes
      - Mitochondria and Chloroplasts
- Passage of Molecules Across Membranes
  - Simple Diffusion
  - Osmosis
  - Facilitated Diffusion
  - Active Transport
  - Group Translocation
  - Engulfment
- Summary

Key Terms

organelle
nucleoid
envelope
appendage
Gram-negative bacterium
Gram-positive bacterium
mycoplasmas
phagocytosis
lipopolysaccharide
lipid
porin
lipoprotein
binding protein
periplasm
cell wall
peptidoglycan
murein
bacillus
coccus
spirillum
vibrio
transpeptidase
lysozyme
turgor
cytoplasmic membrane
transmembrane protein
fluid mosaic model
pilus
pilin
fimbria
adhesin
flagellum
flagellin
monotrichous
amphitrichous
lophotrichous
peritrichous
chemotaxis
aerotaxis
phototaxis
magnetotaxis
axial filament
cytoplasm
ribosome

endospore
sporulation
germination
mitochondrion
chloroplast
cytoskeleton
nucleus
gamete
endoplasmic reticulum
Golgi apparatus
simple diffusion
osmosis
facilitated diffusion
active transport
group translocation
endocytosis
phagocytosis
pinocytosis
exocytosis

Study Tips

1. Have you discovered some study techniques that work best for you? Does scanning the chapter for major topics and subtopics help? Have you tried outlining the chapter content in your own words? Consider writing definitions for key terms in your own words if you have found mastery of vocabulary a major challenge.

2. Answer the several kinds of questions at the end of this chapter and other chapters to test your mastery of the chapter content.

3. Sketch a typical prokaryotic cell. Label the major parts and describe the function of each.

4. Sketch a typical eukaryotic cell. Label the major parts and describe a function of each.

5. You can organize some of the information from questions #3 and #4 another way. List the cell parts and write their functions. Use this format.

Cell Structure	Function

6. Test your understanding of cell transport by answering questions to the following situation. A eukaryotic animal cell has a solute concentration inside of 1.0% with 99% water. The outside environment is 0.5% solutes and 99.5% water. The cytoplasmic membrane of the cell is permeable mainly to the water and not to the solutes.

With reference to the cell which regions are hypertonic and hypotonic?

In which direction will water diffuse in this situation?

How is osmosis demonstrated in this situation?

How will the animal cell change volume over time?

Would a plant cell or bacterial cell experience the same changes as the animal cell in this situation?

The Prokaryotic Cell

A. Label each one of the following statements as true or false, If false, correct it.

1. The term "prokaryotic" means true nucleus.
2. Prokaryotic cells are usually larger than eukaryotic cells.
3. Gram-negative bacteria have a complete cell envelope with all three layers.
4. *Streptococcus mutans* is the bacterium mainly responsible for tooth decay.
5. Gram-positive bacteria are usually more resistant to antibiotics than Gram-negative bacteria.
6. In Gram-negative bacteria the cell wall lies in the periplasm.
7. Teichoic acids add to the integrity of Gram-positive cell walls.
8. Bacilli always divide in only one plane.
9. Turgor pressure develops in a bacterial cell as it loses water.
10. Mycoplasmas have a complex cell wall.
11. The hydrophilic tail of a phospholipid molecule consists of two fatty acid chains.
12. The function of the bacterial cell pilus is mainly locomotion.
13. A bacterium avoids an area with high oxygen by aerotaxis.
14. Eukaryotic ribosomes are 70S.
15. Endospores have a high water content.

B. Match each of the following descriptions to the correct cell structure.

1. capsule	A. can have a peritrichous arrangement
2. cell wall	B. composed of murein
3. endospore	C. appendage for cell attachment
4. flagellum	D. has proteins called porins
5. nucleoid	E. DNA area without a defined membrane
6. outer membrane	F. 70S in bacteria
7. pilus	G. outer, slimy layer of cell envelope
8. ribosome	H. structure resistant to heat

C. Complete each of the following statements with the correct term or terms.

1. The small size of cells permits a _________ growth rate.
2. Smaller cells have a higher surface-to-________ ratio than larger cells.
3. Gram-_________ bacteria lack the outer membrane of the cell envelope.
4. The main function of the cell capsule is ________.
5. The lipopolysaccharide of the outer membrane has a hydrophilic ________ and a hydrophobic ________.
6. _________ are bacteria that lack a cell wall.
7. ________ and _______ are the two sugars with alternating units in the peptidoglycan of the bacterial cell wall.
8. The _________ is the sphere-shaped bacterial cell.
9. Autolysins are _________ that break the cross-linking bonds in peptidoglycan molecules.
10. The ________ of a bacterial cell withstands turgor pressure, preventing the bursting of the cell.
11. Mycoplasmas do not burst by pumping _________ out of their cells.
12. Normal bacteria can be converted to L forms by treating the normal cells with the enzyme ________.

13. The fluid mosaic model refers to the __________ of the bacterial cell.

14. Another word for pilus is ________.

15. A species of the genus __________ causes gonorrhea.

16. Peritrichous refers to an arrangement of _________ around a bacterial cell.

17. By __________ bacteria swim to a region of maximum light intensity.

18. The cytoplasm of the bacterial cell is about 90 percent ________.

19. Antibiotics do not harm the _________ ribosomes of eukaryotic cells.

20. Endospores can germinate to new __________.

The Eukaryotic Cell

A. Label each one of the following statements as true or false. If false, correct it.

1. Bacterial cells are eukaryotic.

2. Eukaryotic flagella produce locomotion by a rapid, rotating motion.

3. The cells of fungi lack a cell wall.

4. The cell walls of plants consist of cellulose.

5. Microtubules are the largest threadlike proteins that compose the cytoskeleton.

6. The nuclear envelope is defined by a single membrane.

7. Eukaryotic DNA is chemically different from prokaryotic DNA.

8. Mitosis is the kind of cell division that produces gametes.

9. The smooth ER is covered with ribosomes.

10. Lysosomes package materials in eukaryotic cells.

B. Match each of the following structures to its correct description.

1. cell wall	A. composes of three threadlike proteins
2. chloroplast	B. contains thylakoids

3.	cilium	C.	repackages new vesicles
4.	cytoskeleton	D.	synthesizes phospholipids
5.	Golgi apparatus	E.	makes proteins
6.	lysosome	F.	resists turgor pressure
7.	RER	G.	has enzymes for digesting substances
8.	SER	H.	cell appendance for locomotion

C. Label each one of the following statements as describing a prokaryotic or eukaryotic cell.

1. The cell contains many kinds of organelles.
2. Human cells are this type.
3. The cell diameter is one to three micrometers.
4. Nuclear envelope is absent.
5. Ribosomes are 70s.
6. Ribosomes are 80s.
7. Cell undergoes meiosis.
8. Proteins for energy release are found in the cytoplasmic membrane.
9. The bacterial cell is this type.
10. Mitochondria and the ER are present.

Passage of Materials Across Cell Membranes

A. Complete each one of the following statements with the correct term or terms.

1. __________ molecules are not soluble in water and pass through the cytoplasmic membrane by dissolving through it.
2. Water passes through the cytoplasmic membrane by passing through ________ in the membrane.

3. _________ is reached if an equal distribution of molecules is established on both sides of a membrane.

4. The cell does not expend ________ for the simple diffusion of molecules across cytoplasmic membranes.

5. _________ is the diffusion of water across a membrane.

6. A ________ environment has a greater concentration of solutes compared to another environment.

7. By ________ diffusion molecules are transported from a region of higher concentration to a region of lower concentration. Carrier molecules mediate this process.

8. By ________ ________ molecules are located from a region of lower concentration to a region of higher concentration. The cell spends energy for this process.

9. ________ ________ is a variation of active transport conducted only by bacterial cells.

10. By ________ solid material is engulfed by a cell.

11. By ________ liquid material is taken in by the cell.

12. ________ is the expulsion of material from the cell.

Discussion Questions

1. Many of the principles of this chapter are used for the physical and chemical control of bacteria. You will learn more about these methods of control in future chapters. Look ahead at these chapters now. Why, for example, does the boiling of materials not kill the cells of species in *Bacillus* and *Clostridium*? How does the use of an autoclave overcome this problem? How do antibiotics such as penicillin or erythromycin inactivate key steps of bacterial metabolism?

2. What evidence do you think supports the idea the eukaryotic cells evolved from prokaryotic cells?

Multiple Choice: Review

1. Select the correct statement about prokaryotic cells.

   A. Human cells are prokaryotic cells.
   B. Most are smaller than eukaryotic cells.
   C. Their nucleus has a defined membrane.
   D. They contain many kinds of organelles.

2. Pneumonia in humans is caused by a species of

   A. *Bacillus.*
   B. *Escherichia.*
   C. *Pseudomonas.*
   D. *Streptococcus.*

3. The ________ is the rod shape of bacterial cells.

   A. bacillus
   B. coccus
   C. spirillum
   D. vibrio

4. A bacteria cell that is hypertonic to its outside environment will

   A. attract water and develop a turgor pressure.
   B. attract water and lose a turgor pressure.
   C. lose water and develop a turgor pressure.
   D. lose water and lose a turgor pressure.

5. By aerotaxis bacteria swim toward a

   A. favorable source of oxygen.
   B. magentic field.
   C. source of glucose.
   D. source of light.

6. Which cell appendage is used for attachment by bacterial cells?

   A. cilium
   B. flagellum
   C. pilus
   D. pseudopodium

7. The 70s-80s ribosome difference accounts for the

   A. action of some antibiotics.
   B. development of a turgor pressure.
   C. differences in the rate of cell division.
   D. storage ability of cells.

8. Which cell organelle conducts protein synthesis?

   A. lysosome
   B. Golgi complex
   C. mitochondrion
   D. RER

9. Each of the following describes simple diffusion except

   A. develops from the random movement of molecules
   B. energy must be spend by cells for it to occur
   C. molecules spread out
   D. tends to produce an equilibrium

10. Osmosis is the movement of

   A. salts through a membrane.
   B. salts without a membrane.
   C. water through a membrane.
   D. water without a membrane.

Answers

The Prokaryotic Cell

A. 1. False: The term "prokaryotic" means before a nucleus. The term "eukaryotic" means true nucleus.
   2. False; A eukaryotic cell is usually about ten times larger than a prokaryotic cell.
   3. True
   4. True
   5. False; Gram-negative bacteria are usually more resistant to antibiotics compared to Gram-positive bacteria.
   6. True
   7. False; Teichoic acids add to the integrity of the Gram-negative cell walls.
   8. True
   9. False: Turgor pressure builds up in a bacterial cell as it gains water.
   10. False; Mycoplasmas are the only group of bacteria that lack a cell wall.
   11. True
   12. False; The function of the pilus for the bacterial cell is usually attachment.
   13. False; By aerotaxis a bacterium swims toward a source of oxygen.
   14. False; Eukaryotic ribosomes are 80s. Prokaryotic ribosomes are 70s.
   15. False; The water content of the bacterial endospore is less than fifteen percent.

B. 1. G 2. B 3. H 4. A 5. E 6. D 7. C 8. F

C. 1. rapid 2. volume 3. positive 4. protection 5. hydrophilic head, hydrophobic tail 6. mycoplasmas 7. NAG, NAM 8. coccus 9. enzymes 10. cell wall 11. sodium ions 12. lysozyme 13. cytoplasmic membrane 14. fimbria 15. *Neisseria* 16. flagella 17. phototaxis 18. water 19. 80s 20. vegetative cells

The Eukaryotic Cell

A.
1. False; Cells of the bacteria and archaea are prokaryotic. All other cells are eukaryotic.
2. False; Eukaryotic flagella produce motion by a slow, undulating motion.
3. False: Fungal cells have a cell wall.
4. True
5. True
6. False; The nuclear envelope is defined by a double membrane.
7. False; Eukaryotic and prokaryotic DNA are identical.
8. False; Mitosis produces the body cells. Meiosis produces gametes.
9. False; The rough ER is covered with ribosomes.
10. False; The Golgi apparatus packages materials in the eukaryotic cell. The lysosome releases enzymes to break down substances.

B. 1. F 2. B 3. H 4. A 5. C 6. G 7. E 8. D

C. 1. eukaryotic 2. eukaryotic 3. prokaryotic 4. prokaryotic 5. prokaryotic 6. eukaryotic 7. eukaryotic 8. prokaryotic 9. prokaryotic 10. eukaryotic

Passage of Materials Across Cell Membranes

A. 1. hydrophobic 2. pores 3. equilibrium 4. energy 5. osmosis 6. hypertonic 7. facilitated 8. active transport 9. group translocation 10. phagocytosis 11. pinocytosis 12. exocytosis

Discussion Questions

1. Study the response of some species of bacteria to intense heat. Study the effects of antibiotics on cell wall or protein synthesis.

2. Compare the structure of prokaryotic and eukaryotic cells. Outline the endosymbiont hypothesis.

Multiple Choice: Review

1. B 2. D 3. A 4. A 5. A 6. C 7. A 8. D 9. B 10. C

Chapter 5
Metabolism of Microorganisms

Metabolism: An Overview
Aerobic Metabolism
- Catabolism
  - Precursor Molecules
  - Reducing Power
  - ATP: Stored Energy
    - ATP Formation
    - Substrate Level Phosphorylation
    - Chemiosmosis
  - Catabolic Pathways
    - Glycolysis
    - The TCA Cycle
    - Pentose Phosphate Pathway
- Biosynthesis
- Polymerization
- Assembly

Anaerobic Metabolism
- Anaerobic Respiration
- Fermentation

Nutritional Classes of Microorganisms
- Formation of Precursor Metabolites by Autotrophs
- Formation of ATP and Reducing Power by Photoautotrophs
  - Cyclic Photophosphorylation
  - Noncyclic Photophosphorylation
- Formation of ATP and Reducing Power by Chemoautotrophs

Regulation of Metabolism
- Purpose
- Types
  - End-Product Inhibition
  - Allosteric Inhibition

Summary

Key Terms

metabolism
catabolism
precursor metabolite
ATP
reducing power
biosynthesis
polymerization

Chapter 5

assembly
oxidation
reduction
dehydrogenation reaction
hydrogenation
free energy
substrate level phosphorylation
chemiosmosis
electron transport chain
electron acceptor
aerobic respiration
glycolysis
TCA cycle
pentose phosphate pathway
anaerobic metabolism
fermentation
denitrifier
autotroph
heterotroph
chemotroph
phototroph
cyclic photophosphorylation
noncyclic photophosphorylation
allosteric inhibition
end-product inhibition

Study Tips

1. Answer the several kinds of questions at the end of this chapter, and other chapters, to test your mastery of key facts and concepts.

2. Outline the broad patterns of glycolysis, aerobic respiration, and photosynthesis from the details of the chapter. Focus on the formation of ATP molecules within your outline.

3. Can you relate the lessons of this chapter to your current experiences in lab? For example, if a bacterium ferments glucose in a Durham tube, what produces the color change in the growth medium from red to yellow? What is the source of gas accumulation in the inverted tube of this setup?

4. Bacteria are excellent models to study patterns of metabolism that apply to other kinds of organisms. Can you find some examples of this in the chapter? Begin with lactic fermentation is animal tissues. What is the source of fatigue in skeletal muscles? What is meant by the oxygen debt that must be repaid after intense exercise by the human body?

Metabolism: An Overview

A. Complete each of the following statements with the correct term.

1. __________ refers to all biochemical reactions that take place in a cell.

2. The number of biochemical reactions occurring in the cell of *E. coli* is about ________.

3. Only twelve organic compounds, ________ metabolites, are needed as starting points for the metabolic reactions in *E. coli*.

4. Through ________ the cell makes small molecules from precursor metabolites.

5. Polymers such as DNA and proteins are the large molecules or__________ of the cells.

6. By the process of________ macromolecules are used to make cell organelles.

Aerobic Metabolism

A. Label each of the following statements as true or false. If false, correct it.

1. Porins are small, water-filled holes in the outer membrane of *E. coli*.

2. Most molecules pass through porins by active transport.

3. Transporters are enzymes that bring nutrients into the cell.

4. Substances enter the bacterial cell by translocation through simple diffusion.

5. Normally *E. coli* transports fructose-6-phosphate into the cell and converts it to glucose-6-phosphate.

6. Reduction is the loss of electrons.

7. During oxidation a proton is usually transported along with an electron.

8. The cell uses its reducing power to make ATP.

9. The reduced form of NAD(P) is NAD(P)+.

10. More stable reactant bonds and more instable product bonds drive a chemical reaction.

11. Most phosphate-containing metabolic intermediates contain high-energy bonds to make ATP.

12. During chemiosmosis the cell uses a proton gradient to make ATP from ADP and phosphate.

13. Cytochromes of the electron transport chain accept electrons and hydrogen ions.

14. Eukaryotic organisms generate six ATP molecules for each pair of electrons passing through the electron transport chain.

15. Central metabolism begins with the sugar sucrose.

16. By glycolysis two molecules of glucose are produced from one molecule of pyruvate.

17. During glycolysis, there is a net gain of six ATP molecules per glucose molecule without the contribution of substrate level phosphorylation.

18. Acetyl CoA enters the TCA by combining with a six-carbon molecule, oxaloacetate.

19. Production of NADH and NADPH increases the reducing power in a cell.

20. *E. coli* produces about 10,000 different enzymes.

21. *E. coli* can make all 20 amino acids required to build proteins.

22. The ordering of amino acids in a protein is determined by the information in the structure of DNA.

23. To make glycogen in *E. coli* , the glucose building block is first converted to glucose-6-phosphate.

24. Glucose is a polymer made from glycogen.

25. Peptidoglycan is the major macromolecule of the outer membrane of the bacterial cell.

Anaerobic Metabolism

A. Complete each of the following statements with the correct term.

1. ________ anaerobes are capable of both aerobic and anaerobic metabolism.

2. In anaerobic respiration a substance other than ________ is the final electron acceptor.

3. ________ are organisms that convert nitrates intro nitrogen gas.

4. Lactic acid bacteria form __(number)__ molecules of ATP per glucose molecule fermented.

5. Deprived of oxygen, the muscle tissue of animals carry out lactic acid ________.

6. In alcoholic fermentation pyruvate is converted to carbon dioxide and ________.

Nutritional Classes of Microorganisms

A. Matching - Match each term to its correct description.

1. autotroph	A. uses inorganic compounds for energy
2. chemotroph	B. light feeder
3. heterotroph	C. self-feeder
4. phototroph	D. different-feeder

B. Complete each of the following statements with the correct term.

1. Autotrophs make precursor metabolites from the gas ________.
2. Autotrophs use the pathways of the ________ cycle to make precursor metabolites.
3. Autotrophs combine carbon dioxide with a five-carbon sugar to produce the substance _________.

C. Label each of the following statements as describing cyclic or noncyclic photophosphorylation.

1. Electrons are rejoined to chlorophyll in its ground state.
2. It does not return electrons to chlorophyll molecules.
3. It is carried out nonoxygenic organisms.
4. It uses photosystem II.
5. It cannot generate reducing power.

Regulation of Metabolism

A. Define each of the following.

1. allosteric enzyme
2. effector
3. end-product inhibition

4. allosteric activation

Discussion Questions

1. Look ahead to Chapter 6 on genetics. How do you think that the genetics of microorganisms relates to their metabolic capabilities? In particular study the function of DNA in cellular metabolism.

2. Study the reactants and products of photosynthesis and cellular respiration. Also look at some of the details of their mechanisms. Are there similarities between these two processes? Are there differences?

Multiple Choice: Review

1. Amino acids are bonded into proteins in a cell. This is an example of

   A. assembly.
   B. biosynthesis.
   C. catabolism.
   D. polymerization.

2. Porins in the outer membrane of the bacterial cell are

   A. enzymes that combine with substrate molecules.
   B. substrate molecules requiring molecules.
   C. water molecules that serve as substrates.
   D. water-filled holes in the outer membrane.

3. About _________ chemical reactions in *E. coli* are required to make the twelve metabolites from glucose.

   A. six
   B. twelve
   C. twenty-five
   D. fifty

4. Dehydrogenation reactions involve the loss of

   A. electrons.
   B. protons.
   C. electrons and protons.
   D. neither electrons nor protons.

5. Acetyl CoA enters the TCA after being formed from

   A. glucose.
   B. glucose-6-phosphate.
   C. oxaloacetate.
   D. pyruvate.

6. The concentration of ________ is an indicator of a cell's reducing power.

   A. ADP
   B. glucose
   C. NAD(P)H
   D. oxaloacetate

7. In the electron transport chain the quinones accept

   A. electrons.
   B. hydrogens.
   C. electrons and hydrogens.
   D. oxygen.

8. Which organism is the self-feeder?

   A. autotroph
   B. chemotroph
   C. heterotroph
   D. phototroph

9. Effectors control the function of allosteric enzymes by changing their

   A. concentration.
   B. location in the cell.
   C. pH requirement.
   D. shape.

10. By end-product inhibition, the buildup of an end-product _________ the rate of the first reaction of the associated metabolic pathway.

    A. decreases
    B. increases

Chapter 5

Answers

Metabolism: An Overview

A. 1. metabolism 2. two thousand 3. precursor 4. biosynthesis 5. macromolecules 6. assembly

Aerobic Metabolism

A.
1. True
2. False; Most molecules pass through porins by simple diffusion.
3. True
4. False; Translocation is an energy-requiring process.
5. False; *E. coli* transports glucose into the cell and converts it to glucose-6-phosphate.
6. False; Oxidation is the loss of electrons. Reduction is the gain of electrons.
7. True
8. True
9. False; NAD(P)+ is the oxidized form of NAD(P).
10. False; More instable reactant bonds and more stable product bonds drive a chemical reaction.
11. False; Only a few metabolic intermediates produced by substrate level phosphorylation have the high energy bonds to make ATP.
12. True
13. False; Cytochromes of the electron transport chain accept only electrons.
14. False; Eukaryotic organisms generate three ATP molecules for each pair of electrons passing through the electron transport chain.
15. False; Central metabolism begins with the sugar glucose.
16. False: By glycolysis one molecule of glucose is converted to two molecules of pyruvate.
17. False: During glycolysis there is a net gain of two ATP molecules per glycose molecule without the contribution of substrate level phosphorylation.
18. False; Acetyl CoA enters the TCA by coming with a four-carbon molecule, oxaloacetate.
19. True
20. False; *E. coli* produces about 2,000 different enzymes.
21. True
22. True
23. False; In *E. coli*, glucose is converted to glucose-1-phosphate to make glycogen.
24. False; Glycogen is a polymer made from the building block glucose.
25. False; Peptidoglycan is the major macromolecule of the cell wall of the bacterial cell.

Anaerobic Metabolism

A. 1. facultative 2. oxygen 3. denitrifiers 4. two 5. fermentation 6. ethanol

Nutritional Classes of Microorganisms

A. 1. C 2. A 3. D 4. B

B. 1. carbon dioxide 2. Calvin-Benson 3. ribulosebisphophate

C. 1. cyclic 2. noncyclic 3. noncyclic 4. noncyclic 5. cyclic

Regulation of Metabolism

1. An allosteric enzyme is an enzyme that changes activity when bonded to a signal molecule.

2. The effector is the signal molecule that binds to an allosteric enzyme to change its activity.

3. The end product of a metabolic pathway binds to the first enzyme in the pathway, inhibiting its activity.

4. An increase in the intracellular concentration of an effector activates an allosteric protein

Discussion Questions

1. Some microorganisms, such as *E. coli* have all the genes needed to make needed precursor metabolites from glucose. Other kinds of bacteria have mutations leading to the loss of this ability. Therefore, they are less independent nutritionally.

2. Photosynthesis uses carbon dioxide and water to make sugar and oxygen. Overall, cellular respiration does the exact opposite. Both processes, however, use a proton energy gradient to make ATP.

Multiple Choice: Review

1. D 2. D 3. C 4. C 5. D 6. C 7. B 8. C 9. D 10. A

Chapter 6
The Genetics of Microorganisms

Structure and Function of Genetic Material
- The Structure of DNA
- Reactions of DNA
- Replication of DNA
  - Replication: The Synthesis of DNA
  - The Mechanics of Replication
- Gene Expression: Transcription
- Gene Expression: Translation
  - The Role of RNA
  - The Mechanics of Translation

Regulation of Gene Expression
- Regulation of Transcription
  - Regulatory Proteins that Bind to DNA
  - The LAC Operon
  - Attenuation
- Regulation of Translation
- Global Regulation
- Two-component Regulatory Systems

Changes in a Cell's Genetic Information
- The Genome
  - Plasmids
  - Genotype and Phenotype
- Mutations
  - Incidence of Mutations
  - Chemical Mutagens
  - Physical Mutagens
  - Biological Mutagens
  - The Consequences of Mutations
  - Damage to the Gene Product
  - Essential Gene Products
- Selecting and Identifying Mutants
  - Direct Selection
  - Indirect Selection
  - Brute Strength
  - Site-directed Mutagenesis
- Uses of Mutant Strains
- Genetic Exchange Among Bacteria
  - Transformation
  - Natural Transformation
  - Artificial Transformation
  - Conjugation
  - Transduction

Key Terms

DNA
nucleotide
replication
gene expression
transcription
translation
nucleoside triphosphate
replication fork
semiconservative replication
replication apparatus
antiparallel
DNA polymerase
ligase
mRNA
tRNA
codon
anticodon
mutation
genome
plasmid
genotype
phenotype
mutagen
transposon
direct selection
indirect selection
brute strength
transformation
conjugation
transduction

Study Tips

1. Continue to answer the several kinds of questions at the end of each chapter. After answering them, ask your instructor to post the answers.

2. Remember that good students also write their own questions, particularly as they anticipate possible exam questions. Work with another student in class and test each other with your questions.

3. Some colleges provide take-apart models of the DNA double helix. Ask your instructor if they are available at your school. Using them can help you to understand the three-dimensional makeup of DNA. It can also help you to understand how it replicates.

4. One of the important ideas in this chapter is the flow of genetic information in the cell. For additional practice, try the following. Write out a series of DNA base pairs. Here is one example:

   TCC - TAT - GCA - CCG - AGT

   Answer these questions:

   If this is one-half of the DNA double helix, what is the base sequence of the other half of the DNA molecule? Read the stated bases from left to right.

   From the originally-stated sequence of DNA bases, what is the series of bases transcribed into mRNA?

   From these five mRNA codons, what is the series of amino acids transcribed by this sequence?

   What is the anticodon for each tRNA molecule carrying each amino acid to the ribosome?

5. As a variation of question #4, try the following by working through the process backwards. Write five tRNA anticodons from left to right.

   What is the order of the mRNA codons that will attract these five anticodons at the ribosome?

   What amino acid will be placed by each mRNA codon at the ribosome? Write them from left to right.

   What is the sequence of bases of DNA that will transcribe the five mRNA codons? Write them from left to right.

Structure and Function of Genetic Material

A. Label each one of the following statements as true or false. If false, correct it.

   1. The base in a nucleotide of DNA can be either A, C, G, or T.

   2. The base pairs of DNA are either A-G or C-T.

   3. By translation RNA makes DNA.

   4. The nucleotide of DNA contains two phosphate groups.

5. During DNA replication, the replication forks move in opposite directions.

6. The terminus is where the bubble forms that initiates replication of the circular, bacterial chromosome.

7. The function of DNA ligase is to cleave the two strands of the DNA double helix.

8. The two strands of the DNA double helix are parallel.

9. The three kinds of RNA are messenger, transfer, and ribosomal.

10. During transcription ribonucleoside triphosphates pair with exposes bases on DNA.

11. The promoter is a structural gene on DNA that transcribes mRNA to make a protein.

12. The anticodon is a base triplet on mRNA.

13. Activation occurs when tRNA is linked to an amino acid.

14. A protein produced by translation corresponds to a single gene.

15. AUG is a termination codon for transcription.

B. Complete each of the following statements with the correct term or terms.

1. _________ is the science that studies the heredity of organisms.

2. The nucleotides of DNA consist of three parts: deoxyribose, phosphate, and a _________.

3. The base pairs of DNA are united by _________ bonding.

4. DNA is synthesized from the polymerization of _________, its building blocks.

5. Deoxynucleotides react with ATP to form nucleotide ________.

6. Each strand of the DNA double helix has a three prime end and a _________ prime end.

7. The replication of DNA is not conservative; it is ________.

8. The type of RNA that carries the amino acid to the ribosome is _________.

9. In prokaryotes transcription is guided by the enzyme ________.

10. Transcription begins at a promoter and ends at a site, the ________.

11. The ________ is the base triplet of tRNA that is complementary to the codon on the mRNA.

12. A ________ codon does not encode for any amino acid.

13. The number of amino acids directed by the genetic code is ________.

14. The ________ - ________ sequence is the ribosome binding site.

15. An mRNA with attached ribosomes and proteins is a _______.

Regulation of Gene Expression

A. Label each of the following as describing inducible enzymes, repressible enzymes, or constitutive enzymes.

 1. Many are enzymes of biosynthetic pathways.

 2. Many are needed to metabolize different sugars.

 3. They are produced when a signal molecule is scarce.

 4. The sugar serves as a signal molecule.

 5. The end product of a pathway is the signal molecule.

 6. They are always produced at a constant rate.

B. Label each one of the following statements about the Lac operon as true or false. If false, correct it.

 1. Galactoside permease splits a disaccharide into two monosaccharides.

 2. Beta-galactosidase brings lactose into the cell.

 3. The gene lacI lies within the lac operon.

 4. The lac repressor binds to the lac operator in the absence of lactose.

 5. The effector, allolactose, has a negative effect on the lac repressor.

C. Complete the following statements about attenuation.

 1. Attenuation usually regulates the production of enzymes in __________ pathways.

 2. The __________ operon is an attenuation regulated operon.

 3. A/an _________ loop prevents the formation of an attenuator loop.

4. When more histidine is needed, the cell makes more of the ________ needed to make the amino acid.

5. Histidine regulates the activity of one of the enzymes to make it by end-product _________.

D. Complete the following statements on global regulation and the two-component regulatory system.

1. By catabolite ________, many genes and operons are regulated and coordinated in response to a source of carbon for growth.

2. CAP is a _________ protein.

3. Cyclic _________ is the signal molecule for global regulation.

4. Most studies of global regulation involve the bacterium ____________.

5. A sensor is a protein that detects an environmental signal and transmits it to another protein, the ________ _________.

6. The regulation of __________ in *E. coli* illustrates the principles of two-component regulation.

Changes in Cell's Genetic Information

A. Match each of the following terms to its correct description.

1. beta galactosidase	A. agent causing genetic change
2. conjugation	B. change in cell's genetic makeup
3. genome	C. effect of genes on a cell function
4. genotype	D. enzyme
5. mutagen	E. cell's genetic plan
6. mutation	F. process that transfers R factor
7. phenotype	G. R factor is one kind
8. plasmid	H. sum total of DNA in a cell

B. Label each of the following as describing a chemical mutagen, physical mutagen, or biological mutagen.

1. Hydroxylamine reacts with cytosine.

2. It involves a transposable element.

3. It involves jumping genes.

4. It produces a thymine dimer.

5. It is produced by 5-BU.

6. It is produced by UV light

C. Define each of the following.

1. missense mutation

2. nonsense mutation

3. lethal mutation

4. conditionally expressed mutation

D. Explain the difference between direct selection and indirect selection for identifying mutants.

E. Label each of the following statements as describing transformation, conjugation, or transduction.

1. A virus transfers genes between cells.

2. Hfr cells carry this out.

3. It can be generalized or specialized.

4. It can be natural or artificial.

5. It involves the F plasmid.

6. It is used to map genes on chromosomes.

7. It requires a pilus.

8. Phages with life cycles are involved.

9. Viruses reproduce at the expense of the host cell in the process.

10. DNA is absorbed from the environment.

11. *Streptococcus pneumoniae* has surface enzymes called nucleases that cut bound DNA into fragments for incorporation in cells.

12. A bacteriophage is involved.

Discussion Questions

1. Much of our knowledge about molecular genetics is humans developed from studying the genetics of microorganisms. What advantages do microorganisms offer for study and how can their genetics be applied to humans?

2. Look ahead to Chapter 13 on viruses. Some viruses have the ability of reverse transcription. Their RNA makes DNA. Can you outline some of the basic steps of this process?

Multiple Choice: Review

1. In the DNA double helix the number of thymine bases always equals the number of __________ bases.

   A. adenine
   B. cytosine
   C. guanine
   D. uracil

2. During transcription of DNA a sequence of ATCG will order an RNA base sequence of

   A. ATCG.
   B. GCTA.
   C. GCUA.
   D. UAGC.

3. The function of tRNA is to

   A. build the ribosome.
   B. carrying amino acids to the ribosome.
   C. catalyze nucleotides.
   D. transport nucleotides to DNA.

4. Select the incorrect description about the anticodon.

   A. It is a base triplet.
   B. It is found on mRNA.
   C. It is smaller than mRNA.
   D. It lacks a nucleotide sequence of bases.

5. The role of the repressor in the lac operon is to

   A. bind to the lac operator.
   B. remove lactose from the cell.
   C. stimulate transcription.
   D. transport lactose into the cell.

6. The genome is the

   A. signal for enzyme induction.
   B. signal for enzyme repression.
   C. sum total of all DNA in the cell.
   D. sum total of all RNA in the cell.

7. A nonsense mutation stops translation before a protein product is complete.

   A. True
   B. False

8. UV light is an example of a source of a _________ mutation.

   A. biological
   B. chemical
   C. physical
   D. spontaneous

9. By conjugation

   A. cells reproduce asexually.
   B. DNA replication is stopped.
   C. plasmids are destroyed.
   D. plasmids are transferred to cells.

10. Transduction is carried out by a

    A. alga.
    B. fungus.
    C. protozoan.
    D. virus.

Answers

Structure and Function of Genetic Material

A. 1. True
   2. False; The base pairs of DNA are A-T and G-C.

3. False; By translation RNA makes protein.
4. False; Each nucleotide of DNA contains one phosphate.
5. True
6. False; The terminus is where two completed chromosomes separate after being copied from the original bacterial chromosome.
7. False; The function of DNA ligase is to seal a segment of DNA into a DNA molecule.
8. False; The two strands of the DNA double helix are antiparallel.
9. True
10. True
11. False; The promoter signals the binding of RNA polymerase for transcription.
12. False; The anticodon is a base triplet on tRNA.
13. True
14. True
15. False; AUG is a codon that initiates transcription

B. 1. genetics 2. nitrogen base 3. hydrogen 4. nucleotides 5. triphosphates 6. five prime 7. semiconservative 8. transfer RNA 9. RNA polymerase 10. terminator 11. anticodon 12. nonsense 13. twenty 14. Shine-Dalgarno 15. polysome

Regulation of Gene Expression

A. 1. repressible 2. inducible 3. repressible 4. inducible 5. repressible 6. constitutive

B.
1. False; Galactoside permease brings lactose into the cell.
2. False; Beta-galactosidase splits a disaccharide into two monosaccharides.
3. False; The gene lacI lies outside the lac operon.
4. True
5. True

C. 1. biosynthesis 2. histidine 3. antiterminator 4. enzymes 5. inhibition

D. 1. repression 2. regulatory 3. AMP 4. *E. coli* 5. response regulator 6. chemotaxis

Changes in the Cell's Genetic Structure

A. 1. D 2. F 3. H 4. E 5. A 6. B 7. C 8. G

B. 1. chemical 2. biological 3. biological 4. physical 5. chemical 6. physical

C.
1. It is a change in a codon to one that codes for a different amino acid.
2. It is a change in a codon to one that stops translation.
3. It is a change in DNA that stops DNA replication
4. It is a genetic change rendering a gene product nonfunctional only in certain environments.

D. By direct selection conditions are created that promote the growth of the desired mutant strain. By indirect selection the growth of a desired mutation strain is prevented. A condition is then imposed on growing cells.

E. 1. transduction 2. conjugation 3. tranduction 4. transformation 5. conjugation 6. conjugation 7. conjugation 8. transduction 9. transduction 10. transformation 11. transformation 12. transduction

Discussion Questions

1. The genetic code is universal in all cells. Microorganisms, particularly bacteria, are easy to cultivate and study compared to human cells. There are common patterns of metabolism in eukaryotic and prokaryotic cells.

2. Study Chapter 13 to learn more about viruses. RNA makes DNA that in turn makes RNA. To make this DNA, uracil in RNA must be complementary to adenine in DNA. What unique enzymes do you think must be present?

Multiple Choice: Review

1. A 2. D 3. B 4. B 5. A 6. C 7. A 8. C 9. D 10. D

Chapter 7
Recombinant DNA Technology

- Recombinant DNA Technology
- Gene Cloning
    - Obtaining DNA
    - Splicing Genes into a Cloning Vector
        - Cutting with Restriction Endonucleases
            - Properties of Restriction Endonucleases
            - Following the Reaction
        - Ligation
    - Putting Recombinant DNA into a Host Cell
    - Finding the Right Gene
        - Prior Purification
        - Subsequent Identification
- Hosts for Recombinant DNA
- Applications of Recombinant DNA Technology
    - Making Proteins
    - Amplifying Genes
    - Engineering Organisms
- Summary

Key Terms

recombination
homologous
gene cloning
cloning vector
cell extract
intron
reverse transcriptase
origin of replication
ligation
restriction endonuclease
rotational symmetry
palindromic
cohesive ends
RFLP
prior purification
polymerase chain reaction

Chapter 7

Study Tips

1. Continue to answer the several kinds of questions at the end of each chapter. After answering them, ask your instructor to post the answers. Did you make any mistakes? Talk to your instructor and class members in order to help you improve your understanding of the facts and concepts in the chapter.

2. This chapter reveals the potential to treat many human diseases through the administration of products developed by recombinant DNA technology. From your study can you list any human diseases that can be treated through this technology? Start with the following list of products that are being used. Choose one that you find interesting and learn more about it from library research and the Web. What is the function of this product in the body? What human disease develops in its absence?

   erythropoietin
   clotting factors
   growth hormone
   insulin
   interferon
   surfactant
   tumor necrosis factor
   tPA
   vaccines for hepatitis

3. Every day in the national news there are stories that draw from the principles explained in this chapter. Consider the following. A small amount of DNA, equivalent to one gene, is collected at a crime scene. For analysis through DNA fingerprinting, about 75,000 copies of the gene are necessary. By the PCR, how many cycles of replication from this original gene are necessary to make the analysis?

4. Read the newspapers and national magazines to find other examples of recombinant DNA technology.

Recombinant DNA Technology/Gene Cloning/Hosts for Recombinant DNA

A. Complete each of the following statements with the correct term or terms.

   1. ________ is the process of forming a new combination of genes, natural or artificial.

   2. In eukaryotes, crossing over occurs between ________ chromosomes.

   3. Gene ________ is the process of obtaining a large number of copies of a gene from a single gene.

   4. A cloning _________ is a DNA molecule that a cell will replicate in the process of gene cloning.

   5. About __(number)__ percent of a bacterial cell consists of DNA.

6. ________ are noncoding regions on the chromosome of a eukaryotic cell.

7. Through the enzyme ________ ________, mRNA makes intron-free DNA.

8. By the process of ________ a fragment of DNA is sealed into a chromosome.

9. Restriction endonucleases evolved in bacteria probably to protect them from attack by ________.

10. Taq I is a restriction endonulcease obtained from the organism ________ ________.

11. AGATCT is a DNA base sequence known as ________.

12. DNA has a ________ charge.

13. Fragments of DNA can be separated by gel ________.

14. *E. coli* will take up DNA from a solution containing the salt __________ along with a sudden temperature change.

15. __________ is a process whereby DNA from a virus is used as a cloning vector.

16. By ________ DNA is introduced into cells when electrical impulses destabilize the plasma membrane.

17. hGH is secreted by the ________ gland in humans.

18. 10 kilobases contains __(number)__ base pairs.

19. Cells of the organism __________ ________ were the host first used to make insulin.

20. Human cells tend to glycosylate protein molecules after synthesis, meaning that they add __________ molecules to the proteins.

B. Define each of the following.

1. recombinant molecule -

2. cloning vector -

3. cell extract -

4. reverse transcriptase -

5. prior purification -

6. transfection -

7. shearing -

8. microinjection -

Applications of DNA Technology

A. Label each of the following statements as true or false. If false, correct it.

1. There are 20 to 40 copies of plasmids (pBR type) per cell of *E. coli*.

2. The number of base pairs in the human genome is about 3 million.

3. PCR can produce a pure solution of a single gene.

4. To perform PCR, four different kinds of deoxyribonucleoside triphosphates are required.

5. The enzyme of *Thermus aquaticus*, taq polymerase, is sensitive at very high temperatures.

6. Plants can be genetically engineered for specific purposes.

Discussion Questions

1. Insertion of a gene for gene therapy in many animals works best at the one-cell stage. Why?

2. The use of recombinant DNA technology and genetic engineering carries ethical questions. Can you think of one that may develop in society?

Multiple Choice: Review

1. Select the correct description about cloning.

   A. One cell produces only a few cells.
   B. It cannot reproduce many copies of a gene.
   C. It is used in recombinant DNA technology.
   D. It uses human cells as vectors for bacterial cells.

2. Reverse transcriptase is an enzyme that directs

   A. RNA making DNA.
   B. RNA making protein.
   C. DNA making DNA.
   D. RNA making protein.

3. Introns are

   A. coding DNA regions in the eukaryotic cell.
   B. coding DNA regions in the prokaryotic cell.
   C. noncoding DNA regions in the eukaryotic cell.
   D. noncoding DNA regions in the prokaryotic cell.

4. Select the palindromic sequence.

   A. ATGCGT
   B. ATAGCG
   C. CGGCGG
   D. GCAGAC

5. Which technique is used to seal together the ends of DNA?

   A. cloning
   B. gel electrophoresis
   C. ligation
   D. restriction

6. Which process inserts DNA into cells using a virus vector?

   A. electroporation
   B. microinjection
   C. transfection
   D. transformation

7. The effect of hGH in an organism is to

   A. decrease growth.
   B. decrease water absorption.
   C. increase growth.
   D. increase water absorption.

8. 3 kilobases contain _________ base pairs.

   A. 3000
   B. 30,000
   C. 300,000
   D. 3,000,000

9. The human genome consists of ________ base pairs.

   A. 3 million
   B. 3 billion
   C. 10 million
   D. 10 billion

10. Glycosylation is the

   A. adding of sugars to proteins.
   B. digestion of proteins into amino acids.
   C. fermentation of sugars.
   D. synthesis of DNA from RNA.

Answers

Recombinant DNA Technology/Gene Cloning/Hosts for Recombinant DNA

A. 1. recombination 2. homologous 3. cloning 4. vector 5. three 6. introns 7. reverse transcriptase 8. ligation 9. viruses 10. *Thermus aquaticus* 11. palindromic 12. negative 13. electrophoresis 14. calcium chloride 15. transfection 16. electroporation 17. pituitary 18. 10,000 19. *E. coli* 20. sugar

B. 1. It is a molecule derived from part of one chromosome and part of another.

   2. It is a DNA molecule that a cell will replicate in the cloning process.

   3. It is the liquid content of ruptured cells.

   4. It is an enzyme that uses mRNA to make a complementary strand of DNA.

   5. It is a process whereby a fragment, carrying a gene to be cloned, is purified and spliced into a cloning vector.

   6. It is putting recombinant DNA into a host cell when a virus is the cloning vector.

   7. It is the process of cutting DNA mechanically when preparing it for recombination into a host cell.

8. DNA is inserted directly into some animal cells by a pipette when in a vacuum.

Applications of DNA Technology

A.
1. True
2. False; The number of base pairs in the human genome is about 3 billion.
3. True
4. True
5. False; This enzyme functions best at high temperatures and can withstand repeated cycles of heating.
6. True

Discussion Questions

1. It is easier to change one cell than many cells. Once changed, the change will normally be duplicated every time cells divide.

2. Start with the idea that genetic engineering can be used to treat gene-based diseases. However, who will decide how this technology will be used fairly?

Multiple Choice: Review

1. C 2. A 3. C 4. B 5. C 6. C 7. C 8. A 9. B 10. A

Chapter 8
The Growth of Microorganisms

Populations
The Way Microorganisms Grow
- Doubling Time and Growth Rate
- Exponential Growth
- Phases of Growth
- Continuous Culture of Microorganisms
- Growth of a Colony

What Do Microorganisms Need to Grow
- Nutrition
  - Carbon
  - Oxygen
  - Nitrogen
  - Phosphorus
  - Sulfur
  - Trace Elements
  - Growth Factors
- The Nonnutrient Environment
  - Temperature
  - Hydrostatic Pressure
  - pH
  - Osmotic Strength

Measuring Microbial Growth
- Measuring the Mass of Cells in a Population
  - Dry Weight
  - Turbidity
- Counting the Number of Cells in a Population
  - Total Cell Count
  - Viable Cell Count
    - Plate count
    - Filtration count
    - Most probable number
  - Measuring Metabolic Activity

Summary

Key Terms

binary fission
doubling time
exponential growth
synchronous culture
lag phase

log phase
stationary phase
death phase
inoculum
batch culture
limiting nutrient
chemostat
growth yield
energy source
hydrogen peroxide
superoxide
hydroxyl radical
superoxide dismutase
peroxidase
growth factor
vitamin
thermophile
mesophile
psychrophile
barophile
acidophile
alkaliphile
halophile
centrifugation
turbidity
dessicator
total cell count
viable cell count

Study Tips

1. Try the review, correlation, and essay questions at the end of the chapter. After answering them, discuss the results with other students in your class. Are your answers similar?

2. Good students are skillful at writing their own questions, particularly as they anticipate possible exam questions. Some of the possible questions in this chapter involve calculations. Here are some examples to get you started.

   Beginning with one cell, can you estimate the size of a bacterial population after three hours? Assume synchronous growth and a doubling time of twenty minutes.

   A bacterial population in nutrient broth has a concentration of 40,000 cells per ml. Can you plot out a series of tenfold serial dilutions that reduces this concentration to 40 cells per ml for plating on a nutrient agar surface?

3. Do you see applications of the information in this chapter to your lab experiences? What does turbidity in a test tube of nutrient broth indicate? Why, through a serial dilution, is choosing plates between 30 to 300 colonies a good compromise between speed and accuracy? How can several colonies in a mixed culture be separated into pure cultures?

4. This chapter has many applications to the upcoming lessons on the physical and chemical control of microorganisms. Look at Chapter 9 for examples.

Populations/The Way Microorganisms Grow

A. Label each of the following statements as true or false. If false, correct it.

1. Microbial growth refers to a change in size of an individual cell.

2. Most microbial cells divide through mitosis.

3. A population with a long doubling time grows rapidly.

4. Cells growing in a bacterial culture are usually not growing synchronously.

5. If an exponential culture is moved from a poor medium to a rich one, no lag normally occurs.

6. The inoculum is the cells used to start a culture.

7. Cells growing on the surface of a colony exposed to air are fully anaerobic.

8. *E. coli* is an obligate anaerobe.

B. Complete each of the following statements with the correct term.

1. By the process of ________ a bubble-like growth enlarges and separates from a parent cell.

2. Another name for generation time is ________ time.

3. *E. coli* synthesizes about 30 __________ molecules as it enters the stationary phase.

4. A ________ culture is grown in a closed container.

5. By the formula of 2 raised to the n power (exponent), after seven generations one original cell will have produced __(number)__ cells. Assume synchronous growth.

6. Cells on the edge of a plated colony are usually in the __________ phase of growth.

C. Match each of the following descriptions to the correct growth phase of a microbial population. Each description matches to only one letter. Throughout the match a letter can be used more than once.

1. exponential growth occurs
2. starting, no-growth period
3. cells start to die
4. growth is very slow
5. occurs before exponential growth
6. most rapid growth rate

A. death
B. lag
C. log
D. stationary

D. Describe each of the following.

1. binary fission
2. closer container
3. limiting nutrient
4. chemostat
5. continuous culture
6. colony

What Microorganisms Need to Grow

A. Label each of the following statements as true or false. If false, correct it.

1. Autotrophic microorganisms obtain their carbon from glucose.
2. Chemoheterotrophs obtain their carbon from carbon dioxide.
3. Oxygenases are enzymes that add atmospheric oxygen directly to organic molecules.
4. Nitrogen makes up about fourteen percent of the dry weight of most microorganisms.
5. Heterocysts are cells that are permeable to oxygen.
6. Hydrogen peroxide is removed from compounds by perioxidases.

7. Most microorganisms can use a variety of organic compounds as a source of oxygen.

8. Sulfur is about 5 percent of the dry weight of cells.

9. *Leuconostoc citrovorum* is a lactic acid bacterium.

10. Psychrophiles grow best at very high temperatures.

11. Thermophiles grow best at very low temperatures.

12. *E. coli* is a mesophile.

13. Prokaryotes living in the deep ocean are barophiles.

14. Acidophiles thirve at high pH values.

15. If the concentration of solutes increases outside a bacterial cell, if faces the threat of losing water by osmosis.

B. Complete each of the following statements with the term greater or less.

1. The growth rate of a bacterial population is _________ in the exponential phase compared to the lag phase.

2. Intracellular ATP is _________ in bacterial cells in the death phase compared to the stationary phase.

3. Bacterial cell biosynthesis is __________ in a poor medium compared to a rich medium.

4. The concentration of sulfur in a cell is __________ than the concentration of nitrogen.

5. The temperature range for bacterial cell growth is __________ than the range for eukaryotic cells.

6. Above optimum temperatures, enzyme activity becomes ________ as the temperature continues to increase.

7. The growth of mesophiles is __________ at 37 degrees Celsius compared to 10 degrees Celsius.

8. Proteins have __________ heat sensitivity compared to other kinds of molecules.

9. If the pH is above 7.6 for *E. coli* , the growth of this bacterium is _________.

10. Halophiles have _________ sensitivity to high salt concentrations outside their cells compared to other kinds of cells.

Measuring Microbial Growth

A. Describe how each of the following is used to measure the number of microorganisms.

1. turbidity
2. dry weight
3. metabolic activity
4. direct count
5. plate count

Discussion Questions

1. Is the projected size of a bacterial population, growing exponentially, simply a matter of doubling the cell number over regular time intervals?

2. Are there similarities between the patterns of worldwide population growth and the growth of a bacterial population in a test tube?

Multiple Choice: Review

1. In the human intestine *E. coli* has a doubling time of about _________ hours.

   A. 4
   B. 8
   C. 12
   D. 16

2. Exponential growth of a population occurs in the _________ phase.

   A. death
   B. lag
   C. log
   D. stationary

3. A closed container for microorganism cultivation means that

   A. four stages of population growth do not occur.
   B. nutrients are constantly supplied.
   C. the population is isolated from external influences.
   D. toxins are constantly removed.

4. *E. coli* is a(n)

    A. facultative anaerobe.
    B. microaerophile.
    C. obligate aerobe.
    D. obligate anaerobe.

5. Oxygenases

    A. add oxygen directly to organic molecules.
    B. break down an organic carbon source.
    C. build an organic carbon source.
    D. remove oxygen from organic molecules.

6. Phosphorus makes up about __________ percent of the dry weight of microorganisms.

    A. three
    B. seven
    C. twelve
    D. fifteen

7. A bacterium thrives at 90 degrees Celsius. It is a(n)

    A. anaerophile.
    B. mesophile.
    C. psychrophile.
    D. thermophile.

8. A barophile is a _________ lover.

    A. acid
    B. base
    C. pressure
    D. salt

9. An acidophile grows best at a pH of

    A. 4
    B. 7
    C. 9
    D. 11

10. The spectrophotometer detects the _________ of a culture.

    A. chemical content
    B. location
    C. pH
    D. turbidity

Answers

Populations/The Way Microorganisms Grow

A.
1. False; Microbial growth refers to a change in the size of the population.
2. False; Most microbial cells divide through binary fission.
3. False; A population with a short doubling time grows rapidly.
4. True
5. True
6. True
7. False; Cells growing on the surface of a colony are fully exposed to air and are aerobic.
8. False; *E. coli* is a facultative anaerobe.

B. 1. budding 2. doubling 3. protein 4. batch 5. one hundred and twenty-eight 6. exponential

C. 1. C 2. D 3. A 4. B 5. B 6. C

D.
1. This is how a bacterial cell elongates and divides.
2. This is a container culturing microorganisms where nutrients are not supplied and toxins are not removed.
3. This is the nutrient in scarcest supply for a population, thus setting the growth of a population.
4. This is a device with an input supplying fresh media to a microbial population. There is an output removing cells and toxins from the population.
5. This is a culture with a constant level of cells due to an input of fresh media and an output for aging cells and toxins.
6. A colony is a solid mass of cells on an agar surface, all produced from the same cell.

What Microorganisms Need to Grow

A.
1. False; Autotrophic microorganisms obtain their carbon from carbon dioxide.
2. False; Chemoheterotrophs obtain their carbon from glucose.
3. True
4. True
5. False; Heterocysts are impermeable to oxygen.
6. True
7. True
8. False; Sulfur is about one percent of the dry weight of microorganisms.

9. True
10. False; Psychrophiles grow best at very low temperatures.
11. False; Thermophiles grow best at very high temperatures.
12. True
13. True
14. False; Acidophiles thrive at low, acidic pH values.
15. True

B. 1. greater 2. less 3. greater 4. less 5. greater 6. less 7. greater 8. greater 9. less 10. less

Measuring Microbial Growth

A. 1. As cell numbers in a population increase, the growth medium becomes cloudier or more turbid. A spectrophotometer can measure this turbidity and estimate cell numbers.

2. The dry weight of a culture can be collected by centrifuging that culture. The weight is divided by the weight per cell, yielding the number of cells.

3. The rate of forming metabolic products, for example, is directly related to cell mass.

4. Cells can be counted in a small volume and this, by proportion, can be computed to a larger volume.

5. A sample of a culture is serially diluted and a small volume is plated on an agar growth surface. Each colony grown represents one cell in the original volume.

Discussion Questions

1. Will all of the cells produced continue to live and reproduce? Also, consider the effect of limiting factors and nutrients.

2. Overall, the human population of the world is increasing exponentially. As there is a finite supply of resources and space, geographically some populations are facing limiting factors. What stages of growth, therefore, are they reaching?

Multiple Choice: Review

1. C 2. C 3. C 4. A 5. A 6. A 7. D 8. C 9. A 10. D

Chapter 9
Controlling Microorganisms

- The Way Microorganisms Die
  - Some Useful Terms
  - Death Rate
  - Sterilization
- Physical Controls of Microorganisms
  - Heat
  - Cold
  - Radiation
  - Filtration
  - Drying
  - Osmotic Strength
- Chemical Controls on Microorganisms
  - Selecting a Germicide
  - Testing Germicides
  - Classes of Germicides
    - Phenol
    - Hexachlorophene
    - Alcohols
    - Halogens and Hydrogen Peroxide
    - Heavy Metals
    - Surfactants
    - Alkylating Agents
- Preserving Food
  - Temperature
    - Canning
    - Pasteurization
  - pH
  - Water
  - Chemicals
- Summary

Key Terms

sterilization
disinfection
decontamination
antisepsis
microbiocidal
microbiostatic
decimal reduction time
thermal death point

thermal death time
heat
radiation
photosensitizer
photoreactivation
filtration
drying
osmotic strength
chemotherapeutic agent
germicide
germistat
paper disc method
use-dilution test
phenol
alcohol
halogen
hydrogen peroxide
alkylating agent
formalin
temperature
canning
pasteurization

Study Tips

1. Continue to answer the questions at the end of the chapter. Ask your instructor to post the answers once you have tried the questions. Discuss the results with your classmates.

2. Vocabulary is a major part of this chapter and the other chapters in the text. Try this format to organize the vocabulary.

Term	Definition

   List any term that you find difficult to describe. Study the text and then write a definition in your own words.

3. There are several important mathematical concepts in this chapter. Ask your instructor for some practice problems on computations involving the D-value and serial dilution.

4. Can you relate some of the concepts of this chapter to the lab component of the course? Why are most cultures grown for study at 48 hours? How does the autoclave sterilize equipment in your lab?

The Way Microorganisms Die

A. Complete each of the following statements with the correct term.

1. ________ is a treatment to destroy all microbial life.

2. __________ kills microorganisms on living tissue.

3. If a treatment is microbiostatic it __________ rather than _________ microorganisms.

4. The D-value is the time in minutes required to kill ___________ percent of the cells in a population.

5. By a heat-killing treatment, _________-phase cells are more susceptible to microbial death than cells in the stationary phase.

6. ___________ ____________ is the bacterium causing botulism that is resistant to heat by forming endospores.

7. The TDP is the __________ temperature required to kill all microorganisms in liquid suspension in 10 minutes.

8. The TDT is the _________ time required to kill all microorganisms in a particular liquid at a given temperature.

Physical Controls on Microorganisms

A. Label each of the following statements as true or false. If false, correct it.

1. Ultraviolet light penetrates through an object to kill all microorganisms.

2. The autoclave employs moist heat.

3. The vegetative cells of thermophilic bacteria can withstand prolonged boiling.

4. Low temperatures are sufficient to kill most microorganisms.

5. Long wavelengths of light are lethal to most microorganisms.

6. UV wavelengths are not visible to humans.

7. X rays add electrons to atoms.

8. Ionizing radiation is rarely used for the physical control of microorganisms in the microbiology laboratory.

9. Viruses can normally be removed from liquids by filters in the microbiology lab.

10. Sublimation is the conversion from the gaseous state to the liquid state.

Chemical Controls on Microorganisms

A. Complete each of the following statements with the correct term.

1. A germicide is a chemical that ________ microorganisms.

2. A ____________ is a chemical that inhibits microbial growth.

3. Germicides are classified as having _________, __________, or __________ activity.

4. The phenol _________is the ratio of endpoints between phenol and another germicide.

5. The end point of a germicide uses a dilution of 1:100000 compared to 1:100 for phenol. The germicide is __(number)__ as powerful as phenol.

6. The size of the zone of ___________ around a disc on an agar surface is used in the paper disc method.

7. The first noticeable effect of phenol on microbial cells is its disruption of the cell _________.

8. Phenolics have a __________ group attached to a benzene ring.

9. In a tincture iodine is combined with ________.

10. A __(numbers)__ percent of hydrogen peroxide is a weak antiseptic.

B. Select the one incorrect statement among the choices for each of the following chemical control methods.

1. phenols/phenolics

   A. These substances are denaturing agents.
   B. They are found in certain throat sprays.
   C. Hexachlorophene is currently restricted by prescription.
   D. Hexachlorophene is still used frequently in hospitals.

2. alcohols

   A. These compounds have a hydroxyl group.
   B. Ethanol is an example.
   C. They kill endospores.
   D. They dissolve lipids on the skin.

3. halogens/hydrogen peroxide

   A. Iodine is an example.
   B. They inactivate proteins.
   C. Free chlorine kills microorganisms.
   D. Hydrogen peroxide is formed by catalase.

4. alkylating agents

   A. Ethylene oxide is an alkylating agent.
   B. They remove short carbon chains from proteins.
   C. Formalin is used to embalm.
   D. Low formaldehyde concentrations are used in vaccines.

Preserving Foods

A. Label each of the following statements as true or false. If false, correct it.

   1. Psychrophilic microorganisms can grow in refrigeration.

   2. Most canning procedures do not attempt to remove the endospores of *Clostridium.*

   3. Pasteurization is a special chemical treatment of milk.

   4. Pasteur developed the process to kill lactic acid bacteria in wine to prevent it from spoiling.

   5. A low pH is an effective means of chemical control against many microorganisms.

   6. Salting is a physical control method used most often for fruits.

Discussion Questions

1. Bacteria cannot normally grow in honey and various kinds of canned fruit preserves. Why?

2. Botulism is food poisoning that develops from eating canned vegetables that are not cooked properly. Why is the thorough cooking necessary?

Chapter 9

Multiple Choice: Review

1. Select the term that involves a treatment to destroy all microbial life.

    A. antisepsis
    B. decontamination
    C. disinfection
    D. sterilization

2. The D-value is the time in minutes required to kill ________ percent of the cells in a population.

    A. 90
    B. 75
    C. 50
    D. 10

3. The ________-phase cells of *E. coli* are killed by cold shock.

    A. death
    B. lag
    C. log
    D. stationary

4. Moist heat treatment kills cells by

    A. altering osmotic conditions.
    B. blocking UV light.
    C. dehydrating cells.
    D. denaturing proteins.

5. The TDP is the ________ temperature required to kill all microorganisms in a particular suspension in ten minutes.

    A. highest
    B. lowest

6. An autoclave normally maintains a temperature of ________ degrees Celsius at 15 pounds per square inch.

    A. 85
    B. 100
    C. 121
    D. 155

7. Which kind of bacterium can grow in a refrigerator?

   A. barophile
   B. mesophile
   C. psychrophile
   D. thermophile

8. The most lethal wavelength of UV radiation to microorganisms is at _________ nm.

   A. 50
   B. 150
   C. 265
   D. 420

9. A _________ percent solution of hydrogen peroxide is needed for sterilization.

   A. 1 to 3
   B. 3 to 6
   C. 6 to 30
   D. 40 to 50

10. Select the incorrect association.

   A. alcohol/ethanol
   B. alkylating agent/ethylene oxide
   C. enzyme/catalase
   D. halogen/hydrogen peroxide

Answers

The Way Microorganisms Die

A. 1. sterilization 2. antisepsis 3. inhibits rather than kills 4. ninety 5. log 6. *Clostridium botulinum.* 7. lowest 8. minimum

Physical Controls on Microorganisms

A. 1. False; UV light only affects the surface.
   2. True
   3. True
   4. False; A low temperature slows down metabolism of most microorganisms but usually does not kill them.
   5. False; Short wavelengths of light can kill microbes.
   6. True

7. False; X rays remove electrons from atoms.
8. True
9. False; Viruses are too small to be removed by the filters.
10. False; Sublimination is the conversion from a solid to a gas.

Chemical Controls on Microorganisms

A. 1. kills 2. germistat 3. high, intermediate, low 4. coefficient 5. one thousand 6. inhibition 7. membrane 8. hydroxyl 9. alcohol 10. three to six

B. 1. D 2. C 3. D 4. B

Preserving Food

A.
1. True
2. False; Canning usually does remove the endospores of *Clostridium botulinum.*
3. False; Pasteurization is the heating of milk for a length of time to remove microorganisms.
4. True
5. True
6. False; Salting is usually employed to preserve fish or meat.

Discussion Questions

1. The honey or preserves represent a hypertonic environment to the microbe cells. How does this affect them osmotically?

2. *Clostridium botulinum* is an obligate anaerobe. Where does it normally live? What disease can it cause? How can prolonged cooking affect it?

Multiple Choice: Review

1. D 2. A 3. C 4. D 5. B 6. C 7. C 8. C 9. B 10. D

Chapter 10
Classification

Principles of Biological Classification
- Scientific Nomenclature
- Artificial and Natural Systems of Classification
- The Fossil Record
- The Concept of Species
  - Bacterial Species
  - Strains
  - Viral Species

Microorganisms and Higher Levels of Classification
Methods of Microbial Classification
- Numerical Taxonomy
- Traditional Characters Used to Classify Bacteria
  - Morphology
  - Biochemistry and Physiology
  - Serology
  - Phage Typing
- Comparing Genomes
  - Percent G + C
  - DNA Hybridization
    - Probes
  - Sequences of Bases in DNA
    - Ribosomal RNA Genes
    - Protein-encoding Genes
  - Characters Used to Classify Viruses
  - Dichotomous Keys

Summary

Key Terms

hierarchy
taxonomy
taxon
specific epithet
genus
species
stromatolite
microbial mats
strain
Eukarya
Bacteria
Archaea

characters
numerical taxonomy
morphology
biochemistry
physiology
serology
phage typing
host range
nosocomial infection
sequencing DNA
DNA hybridization
probe
dichotomous key

Study Tips

1. Answer the several kinds of questions at the end of this chapter. Ask your instructor to check your results. Discuss your answers with the other students in the class.

2. Can you make a dichotomous key? A simple example is sufficient to help you understand the concept of keying. Use a series of geometric shapes for this process: circle, square, rectangle, triangle, hexagon. What character can you use for a first pair of steps that will broadly separate these shapes?

3. How are you progressing in lab? Ask your instructor if you will be identifying an unknown bacterium in this part of the course. Can you list all the Gram-positive and Gram-negative bacteria you have studied to date? Can you use this information as a good first step at keying out an unknown bacterium?

4. What other tests can you use to identify and classify a bacterial unknown? Are there other kinds of stains? Are metabolic tests (e.g., starch hydrolysis, fermentation of glucose) also applicable? Can you further refine the key you started in #3?

Principles of Biological Classification

A. Complete each of the following statements with the correct term.

   1. From species to kingdom, a hierarchical scheme of classification ranks groups that are progressively broader and more _________.

   2. _________ is the science of classification.

   3. The _________ is the lowest taxon where all members are very similar.

4. In *Staphylococcus aureus*, *aureus* is the ___________ name.

5. *E. coli* lives normally in the ________ of the human body.

6. A classification based on phylogeny is a ________ scheme rather than an artificial scheme.

7. The color of the colonies of *Staphylococcus aureus* is ________.

8. __________ are the fossilized remains of phototrophic prokaryotes.

9. The fossil record is limited to the past __(number)__ million years.

10. Groups of bacteria do not always reproduce sexually and share in a common gene _________.

11. The species of a bacterium can be subdivided into _________.

12. The broadest taxon for classifying viruses is the _________.

B. Label each one of the following statements as true or false. If false, correct it.

1. The order is broader and more inclusive as a taxon than the genus.

2. *E. coli* is named in part by where it lives.

3. *Staphylococcus aureus* is named by the arrangement and color of its cells.

4. The Linnaean system of classification is natural.

5. The fossil record is extensive for the last 1.8 million years.

6. Fungi have prokaryotic cells.

7. The strains of *E. coli* are different species of bacteria.

8. Viruses are not classified into kingdoms.

Microorganisns and Higher Levels of Classification

A. Short Answer

1. Why did Haeckel propose a third kingdom, the protists?

2. What type of classification scheme for kingdoms did Whitaker propose?

3. How do the species in Archaea differ from the other bacteria?

4. Why was the Linnaean scheme of two kingdoms found unsatisfactory by most scientists?

5. What are the phyla of eukaryotic organisms?

Methods of Microbial Classification

A. Short Answer

1. What does a high similarity coefficient indicate when classifying microorganisms?

2. What kind of characteristic does the Gram stain provide for bacterial classification?

3. A suspension of bacterial cells is added to tetrazolium dye with a carbon source. The dye changes color. What does this indicate?

4. What kind of molecule is an antibody?

5. How are viruses used in phage typing?

6. If 300 G-C base pairs out of 800 base pairs in a DNA sample, the percentage of A-T base pairs is __(number)__.

7. How is DNA hybridization used for classification?

8. How is the sequencing of DNA used for classification?

Discussion Questions

1. Why has it been difficult to establish a phylogenetic classification for bacteria? Do you think it will become easier?

2. One strain of *E. coli* is a normal floral inhabitant in the human colon. Another form of this same species, if ingested, can poison the body. What do you think makes the two forms different?

Multiple Choice: Review

1. Select the most specific, least inclusive taxon from the following list.

   A. family
   B. genus
   C. kingdom
   D. order

2. In the binomial name *Pseudomonas aeruginosa*, the first name indicates the

   A. family.
   B. genus.
   C. order.
   D. species.

3. The fossil record is limited to about the last __________ million years.

   A. 200
   B. 400
   C. 600
   D. 800

4. Stromatolites are

   A. bacterial life cycles.
   B. competing bacterial imprints.
   C. finely layered rocks.
   D. taxonomic categories.

5. In early classification schemes bacteria were classified as

   A. animals.
   B. fungi.
   C. plants.
   D. protozoa.

6. In early classification schemes protozoan were classified as

   A. animals.
   B. bacteria.
   C. fungi.
   D. plants.

7. The broadest taxonomic category used to classify viruses is the

   A. family.
   B. kingdom.
   C. order.
   D. phylum.

8. Which one of the following is mainly biochemical and physiological characteristic?

   A. cell shape
   B. DNA sequence
   C. habitat preference
   D. pH growth requirement

9. Phage typing uses

   A. bacteria to classify viruses.
   B. viruses to classify bacteria.

10. A DNA sample has 2000 base pairs. 1100 of these base pairs are G-C. The percentage of A-T base pairs is

    A. 40
    B. 45
    C. 55
    D. 65

Answers

Principles of Biological Classification

A. 1. inclusive 2. taxonomy 3. species 4. species 5. colon 6. natural 7. yellow 8. stromatolites 9. 600,000 10. pool 11. strains 12. family

B. 1. True
   2. True
   3. True
   4. False; It is artificial, based on visible similarities.
   5. False; An extensive fossil record exists for the last 600,000 million years.
   6. False; Fungi are eukaryotic.
   7. False; The strains of *E. coli* are genetic variations of the same species.
   8. True

Microorganisms and Higher Levels of Classification

A. 1. Microorganisms cannot be logically classified as plants or animals.

   2. He proposed a five-kingdom scheme. In addition to plants, animals, and protists, he added the kingdoms of monerans (prokaryotes) and fungi.

   3. The bacteria of Archaea are ancient bacteria and are unrelated to the true bacteria.

4. Some species, such as bacteria and protozoa, are not very similar nor phylogenetically related to plants and animals.

5. The phyla of eukaryotes are the animals, plants, fungi and protists.

Methods of Microbial Classification

A.
1. The organisms are closely related at the species or genus level.
2. The Gram stain reveals mainly a morphological characteristic.
3. The dye is reduced and the bacterium can use the carbon source.
4. It is a protein molecule.
5. Phages are used to classify bacteria. If a particular virus lyses a bacterium, there is a clear zone around the bacterium on the lawn of an agar surface. This is a distinctive clue for classification.
6. 500/800 or 62.5%
7. The more hybridization between DNA strands from two organisms, the closer they are related.
8. The greater the similarity between the DNA sequences of two organisms, the more closely they are related.

Discussion Questions

1. Accurate, complete fossil records are rare for bacteria. Why? Restudy the advances in DNA sequencing and explain how this holds promise as a new avenue for classification.

2. The strains are somewhat different genetically. Therefore they make somewhat different products. One is a pathogen, making a toxin that can harm the human body if ingested. The other strain of the bacterium is biochemically compatible with the human body.

Multiple Choice: Review

1. B 2. B 3. C 4. C 5. C 6. A 7. A 8. D 9. B 10. B

Chapter 11
The Prokaryotes

- Prokaryotic Taxonomy
- The Bergey's Manual Scheme of Bacterial Taxonomy
    - The Spirochetes (Section 1)
    - Aerobic, Microaerophilic, Motile, Helical/Fibroid Gram-Negative Bacteria (Section 2)
    - Gram-Negative Rods and Cocci (Section 4)
        - Facultative Anaerobic Gram-Negative Rods (Section 5)
            - The Enterics
            - The Vibrios
            - The Pasteurellas
    - Anaerobic Gram-Negative Straight, Curved, and Helical Rods (Section 6)
    - Dissimilatory Sulfate-and Sulfur-Reducing Bacteria (Section 7)
    - The Rickettsias and Chlamydias (Section 9)
    - The Mycoplasmas (Section 10)
    - The Gram-Positive Cocci (Section 12)
    - Endospore-forming Gram-Positive Rods and Cocci (Section 13)
    - Regular Nonsporing, Gram-Positive Rods (Section 14)
    - Irregular Nonsporing, Gram-Positive Rods (Section 15)
    - The Mycobacteria and Nocardioforms (Sections 16 and 17)
    - Anoxygenic Phototrophic Bacteria (Section 18)
    - Oxygenic Photosynthetic Bacteria (Section 19)
    - Budding and/or Appendaged Bacteria (Section 21)
    - Sheated Bacteria (Section 22)
    - Nonphotosynthetic, Nonfruiting Gliding Bacteria (Section 23)
    - Gliding Fruiting Bacteria (Section 24)
    - Archaeobacteria (Section 25)
        - Methanogens
        - Halophiles
        - Thermoacidophiles
    - Actinomycetes with Mutilocular Sporangia (Section 27)
    - Streptomyces and Related Genera (Section 29)
- Summary

Key Terms

pseudomurein
halophilic
spirochete
rod
coccus
aerobic
anaerobic

microaerophilic
Gram-positive
Gram-negative
crown
transgenic plant
fermentation
butanediol
luminescent
rhinitis
nosocomial
snapping post-fission movement
acid-fast bacterium
purple nonsulfur bacteria
prosthecae
gliding
multilocular

Study Tips

1. Answer the several kinds of questions at the end of the chapter. After answering them, discuss the answers with your instructor and classmates.

2. Can you organize the information in the chapter several ways to help you understand the facts and concepts? Here are some ideas. Can you list the prominent Gram-positive and Gram-negative organisms described in the chapter? Are there other ways to group the bacteria?

3. Ask your instructor to provide some lists of bacteria for keying. Make a dichotomous key for each list. This process will help you for the identification of the bacterial unknown in lab.

4. Can you tie in the examples from this chapter and lecture with your lab experiences? As you observe them under the microscope, which bacteria have independent motility? Which ones are Gram-positive? Which ones are rod-shaped? Which ones have chains of spherical cells? Which ones can ferment lactose?

Prokaryotic Taxonomy

A. Short Answer

   1. What is the recently-developed natural scheme used to classify bacteria?

   2. What is the advantage of using artificial schemes of bacterial classification in standard lab procedures?

   3. List the different properties of bacteria covered in *Bergeys's Manual.*

4. Why is the classification approach used in *Bergey's Manual* more applicable in the microbiology of many undergraduate courses?

The Bergey's Manual Scheme of Bacterial Taxonomy

A. Select the correct description or descriptions for each of the following groups of bacteria.

1. Spirochetes

   A. They are Gram-positive bacteria.
   B. Their axial filaments account for motility.
   C. They are motile in viscous environments.
   D. Some are aerobes.
   E. *E. coli* is a member.

2. Aerobic/Microaerophilic, Motile, Helical/Vibroid Gram-Negative Bacteria

   A. Their flagella are similar to spirochetes.
   B. *Azospirillum* species are anaerobic.
   C. Some species fix nitrogen.
   D. Some strains are widespread in marine environments.

3. Gram-Negative Aerobic Rods and Cocci

   A. *Bordella pertussis* causes whooping cough in children.
   B. *Neisseria gonorrhoeae* is a member.
   C. Xanthan is an antibiotic produced by some species.
   D. *A. tumefaciens* can subvert the metabolism of plant cells.
   E. Some species carry out nitrogen fixation.
   F. Species of *Pseudomonas* contaminate the soil.

4. The Enterics

   A. They are facultatively, anaerobic, Gram-negative rods.
   B. Some are motile by flagella.
   C. All are rod-shaped.
   D. *E. coli* cannot ferment lactose.
   E. *E. coli* cannot change tryptophan to indole.

5. The Vibrios

   A. The shape of their cells is spherical.
   B. Two genera contain luminescent species.

6. The Pasteurellas

   A. Some species cause diseases in humans.
   B. The species of *Haemophilus* produces alcohol commercially.

7. Anaerobic, Gram-Negative, Straight, Curved, and Helical Rods.

   A. They are normally abundant in the blood.
   B. Most species are pathogens.
   C. *Fusobacterium periodonticum* causes dental abscesses.
   D. Large numbers inhabit the human colon.

8. Dissimilatory Sulfate or Sulfur-Reducing Bacteria

   A. They produce the gas hydrogen sulfide.
   B. The group contains rods but not spheres.
   C. One species is responsible for the color of the Black Sea.

9. Anaerobic Gram-Negative Cocci

   A. *Veillonella* is the largest genus.
   B. One species predominates in dental plaque.

10. Rickettsias and Chlamydias

    A. Their cell size is larger than most bacteria.
    B. These groups lack pathogens.
    C. Rickettsiae reproduce by simple division.
    D. Both groups are energy parasites.

11. Mycoplasmas

    A. The members have an extensive cell wall.
    B. All are parasites on humans, animals, or plants.
    C. They can pump sodium ions out of their cells.
    D. Many mycoplasmas cause rhinitis in animals.

12. Gram-Positive Cocci

    A. The species vary considerably in morphology.
    B. The species of *Streptococcus* are strict anaerobes.
    C. Three genera constitute the lactic acid bacteria.
    D. One species causes strep throat.

13. Endospore-Forming Gram-Positive Rods and Cocci

    A. All species produce endospores.
    B. The *Bacillus* species cannot conduct aerobic respiration.
    C. One member of *Clostridium* causes anthrax.
    D. One member of *Bacillus* causes botulism.

14. Regular Nonsporing, Gram-positive Rods

    A. It includes species of *Lactobacillus*.
    B. One species causes the disease listeriosis.

15. Irregular Nonsporing, Gram-Positive Rods

    A. Their species cannot live in the soil.
    B. *Arthrobacter* can move by snapping post-fission movement.
    C. *Bifidobacteria* are anaerobes that ferment lactose.
    D. One pathogen causes acne in humans.

16. Mycobacteria and Nocardioforms

    A. The mycolic acids of two genera have long carbon chains.
    B. One species of *Nocardia* causes tuberculosis.
    C. Its species can be identified by the acid-fast stain.
    D. They are easy to culture.

17. Anoxygenic Phototrophic Bacteria

    A. They use hydrogen sulfide as a source of electrons in photosynthesis.
    B. They are the purple bacteria and green bacteria.
    C. They are found exclusively in anaerobic environments.
    D. Often these bacteria are highly colored.

18. Oxygenic Photosynthetic Bacteria

    A. They are Gram-positive.
    B. Some can fix atmospheric nitrogen.
    C. They resist dessication well.
    D. They have extensive mineral requirements to grow.

19. Aerobic Chemolithotropic Bacteria and Associated Organisms

    A. They obtain energy by oxidizing organic compounds.
    B. Most play a role in the nitrogen cycle.

20. Budding and/or Appendaged Bacteria

    A. The prosthecate bacteria have appendages.
    B. The prosthecae are reproductive structures.
    C. *Caulobacter* has an elaborate life cycle.
    D. *Hyphomicrobium* carries out budding.

21. Sheathed Bacteria

    A. The sheaths are protein tubes.
    B. They are found in contaminated streams.
    C. Some can oxidize iron.

22. Nonphotosynthetic, Nonfruiting Gliding Bacteria

    A. They can only glide in water.
    B. The gliding is a selective advantage.
    C. *Cytophaga* breaks down cellulose in plants.

23. Gliding Fruiting Bacteria

    A. They are soil organisms.
    B. Fruiting bodies can be seen only through a microscope.

24. Archaeobacteria

    A. They are evolutionarily different from other bacteria.
    B. They are divided into three subgroups.
    C. Some species make natural gas.
    D. The halophiles cannot tolerate high salt concentrations.
    E. Bacteriorhodopsin is an enzyme that makes ATP.
    F. Thermoacidophiles are acid loving and heat loving.

25. Actinomycetes with Multilocular Sporangia

    A. The species are halophiles.
    B. *Dermatophilius* infects the human intestine.
    C. *Frankia* fixes nitrogen.

26. Streptomyces and Related Genera

    A. Most members are abundant in the soil.
    B. Their colonies are colorless.
    C. In cross-section their cells appear to be Gram- negative.

Chapter 11

Discussion Questions

1. This chapter consolidates most of the topics of microbiology that you have learned to date. What are these topics and how do they apply?

2. Do you see how the content of lecture and lab are inseparable for microbiology? How does the content of this chapter prove this concept?

Multiple Choice: Review

1. A natural classification scheme for bacteria relies mainly on

   A. cell morphology.
   B. DNA sequencing.
   C. Gram-staining.
   D. motility studies.

2. Select the incorrect characteristic about spirochetes.

   A. Their axial filaments produce unusual motility.
   B. They are Gram-positive.
   C. They are long and helical.
   D. They move rapidly under the microscope.

3. *Rhizobium* is important for

   A. genetic engineering.
   B. nitrogen fixation.
   C. wine fermentation.
   D. yogurt production.

4. Species of *Salmonella*, *Shigella*, and *Yersinia* have the common characteristic of

   A. carrying out nitrogen fixation.
   B. fermenting wine.
   C. lacking a vector as part of their life cycle.
   D. living as enteric bacteria.

5. Vibrios are

   A. curved rods.
   B. rods.
   C. spheres.
   D. spirochetes.

6. Species of *Staphylococcus* are Gram-

   A. positive.
   B. negative.

7. *Bacillus* is a

   A. Gram-negative rod.
   B. Gram-negative sphere.
   C. Gram-positive rod.
   D. Gram-positive sphere.

8. The cyanobacteria were once called the

   A. blue-green algae.
   B. green-algae.
   C. red algae.
   D. yellow-brown algae.

9. Which genus has a species causing tuberculosis?

   A. *Bacillus*
   B. *Escherichia*
   C. *Mycobacterium*
   D. *Staphylococcus*

10. Each is a subgroup of the archaeobacteria except the

    A. halophiles.
    B. methanogens.
    C. sulfur-producers.
    D. thermoacidophiles.

Answers

Prokaryotic Taxomony

A. 1. This scheme uses DNA sequencing. It measures the degree of relatedness of two organisms and emphasizes a phylogenetic basis.

   2. Artificial schemes of bacterial classification are more practical. They are based on visible similarities often observed in the lab.

   3. The properties include physiology, ecology, cultivation, preservation, identification, classification, motility, and cell morphology.

4. It does not emphasize evolutionary relationships and uses more practical means of classification. Characteristics encountered in the lab are featured.

The Bergey's Manual Scheme of Bacterial Taxonomy

A.
1. B, C, D
2. C, D
3. A, B, D, E
4. A, B, C
5. B
6. A
7. C, D
8. A, C
9. A, B
10. C, D
11. B, C, D
12. C, D
13. A
14. A, B
15. B, C, D
16. A, C
17. B, C, D
18. B, C
19. B
20. A, C, D
21. B, C
22. B, C
23. A
24. A, B, C, F
25. C
26. A

Discussion Questions

1. Start by reviewing the information in Chapters 3 through 6. Concentrate on characteristics such as cell morphology, staining properties, motility, metabolic capabilities, and genetics.

2. What characteristics have you been studying in lab? Apply these to one bacterium such as *E. coli*. Describe its: cell shape, Gram-stain reaction, motility, fermentation capability, etc.

Multiple Choice: Review

1. B 2. B 3. B 4. D 5. A 6. A 7. C 8. A 9. C 10. C

Chapter 12
Eukaryotic Microorganisms, Helminths, and Arthropod Vectors

- Eukaryotic Microorganisms
  - Fungi
    - Morphology
    - Ecology
    - Reproduction
    - The Lower Fungi
      - *Chytridiomycetes*
      - *Oomycetes*
      - *Zygomycetes*
    - The Higher Fungi
      - *Ascomycetes*
      - *Basidiomycetes*
      - *Deuteromycetes*
    - Yeasts
    - Dimorphism
    - Plant Diseases
    - Human Health
  - Algae
    - Ecology and Uses
    - Classification
    - Phycology
  - Lichens
  - Protozoa
    - Flagellate Protozoa
    - Amoeboid Protozoa
    - Sporozoa
    - Ciliate Protozoa
  - The Slime Molds
    - True Slime Molds
    - Cellular Slime Molds
  - Helminths
    - Flat Worms
      - Tapeworms
      - Flukes
    - Round Worms
  - Arthropod Vectors
    - The Organisms
    - Arthropods and Human Health
- Summary

Key Terms

saprophyte
mycelium
hypha
thallus
sporangium
conidiophore
ascospore
basidiospore
dimorphism
phytoplankton
rhizoid
flagellate
amoeboid
ciliate
trophozoite
sporozoite
merozoite
plasmodium
fruiting body
scolex
proglottid
mechanical vector
biological vector

Study Tips

1. Continue to answer the several kinds of questions at the end of the chapter.

2. Have you developed a study technique that works best for you? Here is one suggestion. Begin by scanning the major ideas in this chapter. Can you compare the fungi, algae, lichens, and slime molds for similarities and differences?

3. Vocabulary remains a major challenge for any student to master in a science course. Begin by listing the key terms down the lefthand side on a sheet of paper. After studying the chapter, write a definition across from each term in your own words.

4. Up to this point, most of your study has been devoted to the bacteria and viruses. This chapter presents new groups of microorganisms. It also presents the helminths and arthropod vectors. Your lab experience can help extend your study of the unique groups of organisms discussed in this chapter.

Chapter 12

Eukayrotic Microorganisms

Fungi

A. Label each one of the following statements as true or false, If false, correct the statement.

1. An green organism is discovered in a forest ecosystem. Analysis of its cells reveals chloroplasts. This organism could be a fungus.

2. The kingdom of fungi does not contain any single-celled organisms.

3. A mycelium is a mass of hyphae.

4. The thallus is the body of a fungus.

5. An organism is observed in a controlled laboratory environment. It grows rapidly in low-oxygen conditions. There is a high probability that this organism is a fungus.

6. The fungi are a major group of decomposers in the environment.

7. Sporangiospores are produced in conidiophores.

8. The cell structure of a microorganism is observed under the microscope. The cells are well-defined and each cell has one nucleus. This organism could be a lower fungus.

9. The oomycetes cannot live in the soil, as water is required for their survival.

10. In the life cycle of *Rhizopus*, a diploid zygospore divides by meiosis to produce a structure that develops into a sporangium.

11. The species of Deuteromycetes form basidiocarps.

12. The dikaryon phase of a mushroom develops underground.

13. A microorganism is studied under the microscope. It consists of long chains of conidia. This microorganism could be a species of the genus *Penicillium*.

14. During the baking of bread, *Saccharomyces cerevisiae* makes the dough of the bread rise through the process of photosynthesis.

15. One species of yeast is a facultative anaerobe. It growth can be inhibited by placing it in a high-oxygen environment.

16. Most plant diseases are caused by fungi.

17. A fungus caused Dutch elm disease.

18. A person suffers from a fungal skin infection. A antifungal, oral medication is the best strategy for effective treatment.

19. Alfatoxin is a fungal toxin.

20. Headaches in humans is caused by a fungus.

Algae

A. Complete each of the following statements with the correct term or terms.

1. Phytoplankton use the gas __________ in the process of photosynthesis.

2. Phytoplankton produce the gas __________ in the process of photosynthesis.

3. Through an infection of the alga with the genus name __________, a person has difficulty walking.

4. From a red alga, the product __________ is used as a milk thickener.

5. __________ earth is used to filter liquids.

6. In the absence of the pigments ________ and ________, chlorophyll cannot receive light to conduct photosynthesis.

7. An alga does not conduct substantial photosynthesis in the presence of __________ wavelengths of light. It does conduct significant photosynthesis in the presence of _________ and __________ wavelengths of light.

8. __________ is the study of algae.

9. Absence of an __________ in *Euglena* will inhibit its ability to detect light.

10. A microorganism cannot be a species of *Chlamydomonas* if its flagellum is on the __________ side of the body.

Lichens

A. Complete each of the following statements with the correct term or terms.

1. A lichen is a symbiotic association between a __________ and a ________.

2. A lichen is studied in lab. Most of its body mass consists of a __________ organism.

3. *Roccella tinctoria* is the source of an indicator that turns from blue to __________ in an acid environment.

4. An algal species in a lichen produces the gas ________.

5. A cyanobacterium in a lichen fixes __________, an element found in proteins but not found in sugars.

Protozoa

A. Match each description to its correct group of protozoans.

1. move by flagella
2. move by pseudopodia
3. adults are nonmotile
4. all are parasites
5. move by short, numerous hairlike projections
6. *Paramecium* is a member
7. contains the amoeboids
8. species have a macronucleus and micronucleus
9. one species causes malaria
10. contains the trypanosomes
11. diplomonads are included
12. sporozoite infects red blood cells
13. merozoite is part of one specie's life cycle
14. *Didinium* attacks a species of this group
15. some species live in the rumen of cattle

A. Ciliophora
B. Mastigophora
C. Sarcodina
D. Sporozoa

Slime Molds

A. Complete each one of the following with the correct term.

1. Species of Myxogastria are the ___________ slime molds.

2. A plasmodium is a ___________ cytoplasm.

3. Spores of Myxogastria are ________, as their spores contain only one set of chromosomes.

4. Species of Acrasieae are the ____________ slime molds.

5. A grex will probably show reduced motility in the absence of ________.

Helminths

A. Flatworms

1. Select the characterisitics that are descriptive of the body of species of Platyhelminthes.

    A. radial symmetry
    B. flattened body
    C. digestive tract present in parasites
    D. nervous system present
    E. excretory system present

2. List five characteristics of the species of Cestoda.

3. What are cystecerci?

4. List three characteristics of the species of Trematoda.

5. Outline the series of larvae of a fluke after it infects the lung of a host.

B. Roundworms

1. Select the characteristics that are not descriptive of the nematode body.

    A. bilateral symmetry
    B. flattened body
    C. complete digestive tract
    D. well-developed reproductive system
    E. have larval and adult forms
    F. larvae do not form cysts in muscles

Chapter 12

Arthropod Vectors

A. Complete each of the following with the correct term or terms.

1. What are some key adaptations accounting for the success of arthropods on the Earth?

2. What is another name for the group of ticks and mites?

3. Why is a study of the arthropods important in microbiology?

4. Name the three major body regions of an insect?

5. By number of species, how do you evaluate the success of arthropods?

6. What is transovarial transmission?

Discussion Questions

1. How do the pathogenic microorganisms discussed in this chapter overcome the mechanisms of the human immune system to cause their effects in the human body?

2. In the course of evolution do you think the fungi are more closely related to plants or animals?

Multiple Choice: Review

1. Select the correct description about the fungi.

   A. None of their species are parasites.
   B. Some form hyphae and mycelia.
   C. They are phototrophic organisms.
   D. They have a prokaryotic cell structure.

2. Among yeasts budding is a means of

   A. energy storage.
   B. feeding.
   C. movement
   D. reproduction.

3. Which class is not a member of the lower fungi?

   A. Ascomycetes
   B. Chytridiomycetes
   C. Oomycetes
   D. Zygomycetes

4. Select the correct statement about the Deuteromycetes.

   A. They are identified by their pigmentation system.
   B. They belong to the lower fungi.
   C. They produce asexual spores.
   D. They produce only conidia.

5. Phycology is the study of

   A. algae.
   B. bacteria.
   C. fungi.
   D. plants.

6. A lichen is an example of a symbiotic association called

   A. commensalism.
   B. mutualism.
   C. parasitism.
   D. predation.

7. Protozoa are classified into groups based on differences in their

   A. cell shape.
   B. means of motility.
   C. means of locomotion.
   D. type of reproduction.

8. A protozoan with flagella is the

   A. amoeba.
   B. didinium.
   C. euglena.
   D. paramecium.

9. Name the free-swimming larva of *Paragonimus westermani* that encysts in crayfish.

   A. cercaria
   B. cystecerci
   C. miracidia
   D. redia

10. A vector

    A. covers the surface of an arthropod body.
    B. is a specialized reproductive cell.
    C. serves as an antigenic substance.
    D. transports a microbe from one host to another.

Answers

Eukaryotic Microorganisms

Fungi

A. 1. False; A fungus is saprophytic.
   2. False; Yeasts are single-celled.
   3. True
   4. True
   5. False; Most fungi are aerobes.
   6. True
   7. False; Sporangiospores are produced on sporangia.
   8. False; Lower fungi are coenocytic.
   9. False; They need water, but a film of water exists around soil particles.
   10. True
   11. False; Species of Basidiomycetes form basidiocarps.
   12. True
   13. True
   14. False; It produces carbon dioxide by fermentation.
   15. False; It can grow in the presence or absence of oxygen.
   16. True
   17. True
   18. False; It is difficult for drugs to enter this body region via the bloodstream.
   19. True
   20. False; Fungal products are used to treat headaches.

Algae

A. 1. carbon dioxide 2. oxygen 3. *Prototheca* 4. carrageenan 5. diatomaceous 6. carotenoids and phycobilins 7. does not in green; does in blue and red; 8. phycology 9. eyespot 10. anterior

Lichens

A. 1. fungus, phototroph; 2. fungal 3. red 4. oxygen 5. nitrogen

Protozoa

A. 1. B 2. C 3. D 4. D 5. A 6. A 7. C 8. A 9. D 10. B 11. B 12. D 13. D 14. A 15. B

The Slime Molds

A. 1. true 2. multinucleate 3. haploid 4. cellular 5. light

Helminths

A. 1. B, D, E
   2. flattened, segmented body; scolex, proglottids, germinal center, hermaphrodite
   3. The are the larvae of *Taenia saginata* embedded in the skeletal muscles of a host.
   4. flattened body, no segments, ventral sucker present
   5. miracidia - redia - cercaria

B. 1. B, F

Arthropod Vectors

A. 1. exoskeleton and jointed appendages
   2. chelicerates
   3. They are vectors for the transmission of microorganisms.
   4. head, thorax, and abdomen
   5. Their are over 800,000 species of insects alone. They have high species diversity. These various species occupy many different ecological niches.
   6. Ticks pass infectious microorganisms into their eggs.

Chapter 12

Discussion Questions

1. Begin by studying the means of cellular and humoral immunity. How, for example, does an infectious protozoan overcome the phagocytosis and/or antibody production of white blood cells?

2. A fungus has a plantlike appearance. However, it is more animal-like by its heterotrophic mode of nutrition.

Multiple Choice: Review

1. B 2. D 3. A 4. D 5. A 6. B 7. B 8. C 9. A 10. D

Chapter 13
The Viruses

The Ultimate Parasites
- The Discovery of Viruses
- Are Viruses Alive?

Classification of Viruses
- Host Range
- Size
- Structure
  - Nucleic Acid
  - Viral Capsids
  - Viral Envelopes
- Life Cycle
- Taxonomy

Bacteriophages
- Phage Counts and Phage Growth
  - Plaque Count
  - The One-Step Growth Curve
- Replication Pathways
  - Virulent Phages
  - Temperate Phages
  - Diversity

Animal Viruses
- Cell Cultures
- Replication
  - Adsorption
  - Penetration
  - Uncoating
  - Viral Synthesis
  - Maturation
  - Release
- Latency
- Animal Viruses of Special Interest
  - HIV/AIDS
  - Influenza Virus
  - Tumor Virus

Plant Viruses
- Growth, Replication, and Control
- Tobacco Mosaic Virus

Viruses of Eukaryotic Microorganisms

Infectious Agents That Are Simpler Than Viruses
- Viroids
- Prions

Summary

Chapter 13

Key Terms

virus
virion
bacteriophage
host range
capsid
envelope
plus-strand
minus-strand
capsomere
icosohedral
adsorption
penetration
uncoating
lawn
circular plaque
plaque-forming unit
burst period
burst size
virion
eclipse period
lytic pathway
lysogenic pathway
virulent
temperate
prophage
primary cell culture
diploid cell line
heteroploid cell line
continuous cell line
maturation
release
retrovirus
antigenic drift
reverse transcriptase
benign
invasive
malignant
oncogene
TMV
mycovirus
viroid
prion

Study Tips

1. Answer the several kinds of questions at the end of the chapter. Discuss the answers with your instructor and classmates.

2. From your study of the chapter and lecture notes, sketch the basic makeup of a virus. Check the text for accuracy after you have made these sketches on your own.

3. Outline the life cycle of the virulent phage and temperate phage. Express these outlines in your own terms rather than simply copying the information from the text. Ask your instructor to evaluate the accuracy of your outlines.

4. Viruses are constantly in the news. Select one recent article that your can find from a library search and/or Web search. Bring that article to class and discuss it with your instructor and classmates.

The Ultimate Parasites

A. Describe the contributions of each of the following scientists to the discovery of viruses.

    1. Iwanowski -

    2. Beijerinck -

    3. Stanley -

B. Summarize the characterisitics of viruses.

Classification of Viruses

A. Label each of the following as true or false. If false, correct it.

    1. Bacteriophages are viruses.

    2. A specific virus usually attacks a variety of species.

    3. The T4 virus contains over 300 genes.

    4. The RNA molecules of RNA viruses never store genetic information.

    5. Minus strand RNA is converted into mRNA after it enters a host cell.

    6. The capsid is the nucleic acid core of the virus.

    7. The most common polyhedral virus is the icosahedral virus.

8. The membrane that surrounds enveloped viruses is a piece of the host cell's membrane.

9. Nonpolar solvents destroy the genes in a virus.

10. By penetration a virus attaches to the host cell membrane.

Bacteriophages

A. Complete each of the following statements with the correct term.

1. The first bacteriophages studied were seven examples that attacked the organism ________.

2. By the plaque count a nutrient agar surface first supports a ________ of host bacterial cells.

3. By the plaque count a plaque is the result of an epidemic started by one _________.

4. In the one-step growth curve there is no viral increase during the _________ period.

5. The _________ size is the average number of new virions released from a cell infected by viruses.

6. __________ phages follow only the lytic pathway.

7. __________ phages follow the lytic or lysogenic pathway.

8. __________ is when the component parts of new virions are assembled near the end of the latent period.

9. The phage DNA integrated into a host chromosome is a ________.

10. Only _________ strains of *C. diphtheriae* cause diptheria.

Animal Viruses

A. Complete each of the following statements with the correct term.

1. In the 1930s virologists learned that some viruses could be cultured in ________ eggs.

2. A ________ is a confluent of animal cells, one layer thick, in a petri dish.

3. ________ cell cultures are started from normal tissues taken directly from humans or other animals.

4. Continuous cell lines are usually derived from ________ tissue.

5. The _________ cell line is the most famous continuous cell line.

B. Describe the events of each of the following stages of viral replication.

1. adsorption -
2. penetration -
3. uncoating -
4. viral synthesis -
5. maturation -
6. release -
7. latency -

C. Match each of the descriptions to the correct animal virus.

1. families of A, B, C	A. HIV/AIDs virus
2. uses a reverse transcriptase	B. influenza virus
3. has NA spikes	C. tumor virus
4. some cause malignant growths	
5. RNA is eight separate pieces	
6. the virion is diploid	
7. retrovirus	
8. has HA spikes	

Plant Viruses

A. Label each of the statements as true or false. If false, correct it.

1. Plant viruses are usually named by their host range.
2. Aphids transmit plant viruses.
3. Plant viruses are never enveloped.

4. Most viral diseases of plants quickly kill the plants.

5. TMV is named by it unique protein coat.

Viruses of Eukaryotic Microorganisms/Infectious Agents That Are Simpler Than Viruses

A. Complete each of the following statements with the correct term.

1. Until __(number)__ years ago, microbiologists did not believe that viruses infected eukaryotic cells.

2. ________ are viruses that infect fungi.

3. A ________ is a circular molecule of ssRNA without a capsid.

4. The spindle tuber viroid is composed of only __(number)__ nucleotides.

5. The term _________ means proteinaceous infectious particles.

The Origin of Viruses

A. 1. Explain an accepted theory among scientists about how viruses evolved.

Discussion Questions

1. What are some examples of health hazards and diseases that viruses have posed to the human population?

2. You are suffering from the common cold. You visit a physician and ask for a treatment of antibiotics. You are told this treatment will not be effective. Why?

Multiple Choice: Review

1. Select the incorrect statement about viruses.

   A. All cellular organisms are attacked by them.
   B. They are parasites.
   C. They contain a nucleic acid inside a protein coat.
   D. They have a cellular makeup.

2. Viruses are classified by each of the following except

   A. host range.
   B. life cycle.
   C. lipid synthesis.
   D. size.

3. The viral nucleic acid enters the host cell. This step of its life cycle is

   A. adsorption.
   B. bursting.
   C. penetration.
   D. release.

4. The broadest taxonomic category applied to viruses is the

   A. class.
   B. family.
   C. kingdom.
   D. phylum.

5. PFUs are

   A. infected cells only.
   B. virions only.
   C. infected cells and virions.
   D. neither infected cells nor virions.

6. The first stage of viral replication is

   A. adsorption.
   B. maturation.
   C. penetration.
   D. release.

7. Virulent phage infections are swift and deadly.

   A. True
   B. False

8. Monolayers can be separated into individual cells for viral subculturing by briefly treating them with

   A. cellulose.
   B. glucose.
   C. sodium ions.
   D. trypsin.

9. Plant viruses are normally named by the

   A. amount of pigmentation.
   B. disease they cause.
   C. host range.
   D. size of their particles.

10. HIV/AIDS is a

   A. influenza virus.
   B. retrovirus.
   C. TMV.
   D. tumor virus.

Answers

The Ultimate Parasites

A. 1. He found that filters that normally removed most microorganisms did not trap the causative agent in the extracts of plants with the tobacco mosaic disease.

   2. He suggested that microorganisms causing the tobacco mosaic disease are smaller than bacteria and must pass through the pores of the filters.

   3. He crystallized the TMV virus, proving its existence.

B. 1. Viruses are intracellular parasites. They consist of a protein coat surrounding a nucleic acid. They can be viewed as bits of genetic information that instruct a host cell to make more viruses. They are small enough to pass through bacteria-proof filters. Some cause disease. Viruses lack a cellular structure.

Classification of Viruses

A. 1. True
   2. False; A specific virus usually attacks only one species.
   3. False; The T4 virus contains 77 genes.
   4. True

5. True
6. False; The capsid is the protein envelope of the virus.
7. True
8. True
9. False; Nonpolar solvents destroy the viral cell membrane.
10. False; By adsorption a virus attaches to a host cell membrane. By penetration its nucleic acid enters the host cell.

Bacteriophages

A. 1. *E. coli* 2. lawn 3. virion 4. latent 5. burst 6. virulent 7. temperate 8. maturation 9. prophage 10. lysogenic

Animal Viruses

A. 1. chicken 2. monolayer 3. primary 4. cancerous 5. HeLa

B. 1. Receptor-binding proteins of the virions bind to cell receptors of the animal cell.

2. Viruses enter host cells by fusing with the cell membrane, by phagocytosis, or by passing through the host cell membrane.

3. The capsid is removed, making the viral genes naked in the animal cells.

4. The viral nucleic acid takes over metabolism of the host cell, using the ribosomes, ATP, and other raw materials to make new virions.

5. Capsids assemble around the nucleic acids of the new virions.

6. The new virions push through the plasma membrane of the host cell and leave the cell.

7. The viral nucleic acid is incorporated into the host chromosome, but does not express itself. It remains latent.

C. 1. B 2. A 3. B 4. C 5. B 6. A 7. A 8. B

Plant Viruses

A. 1. False; They are named by the disease they cause.
2. True
3. True
4. False; They produce slow, degenerative diseases.
5. False; It is named by the spotting pattern it causes on disease.

Chapter 13

Viruses of Eukaryotic Microorganisms/Infectious Agents That Are Simpler Than Viruses

A. 1. forty 2. mycoviruses 3. viroid 4. 357 5. prion

The Origin of Viruses

A. 1. Viruses may have evolved from the cells they originally infected. They left as part of the genome of ancient cells and currently are infecting new cells.

Discussion Questions

1. Examples range from the common cold to several kinds of cancer in humans and AIDS.

2. Most antibiotics inactivate some step of protein synthesis. Viruses lack ribosomes and cannot conduct protein synthesis. The common cold is caused by a virus.

Multiple Choice: Review

1. D 2. C 3. C 4. B 5. C 6. A 7. A 8. D 9. B 10. B

Chapter 14
Microorganisms and Human Health

Normal Biota
- Resident Biota
- Transient Biota
- Opportunists
- Changing Biota

Symbioses
- Commensalism
- Mutualism
- Parasitism

Factors That Determine the Normal Biota
- Living Tissues as an Environment for Microorganisms
- Structural Defenses
- Mechanical Defenses
- Biochemical Defenses
- Microbial Adaptations to Life on Body Surfaces

Sites of Normal Biota
- The Skin
  - Microbial Environment
  - Biota of the Skin
- The Conjunctiva
  - Microbial Environment
  - Biota
- The Nasal Cavity and Nasopharynx
  - Microbial Environment
  - Biota
- The Mouth
  - Microbial Environment
  - Biota
- The Intestinal Tract
  - Microbial Environment
  - Biota
- The Vagina
  - Microbial Environment
  - Biota

Is the Normal Biota Helpful or Harmful
- Conflicting Theories
- Harmful Effects
- Beneficial Effects
- Loss of Normal Biota
- A Spectrum of Disease-Causing Abilities

Summary

Chapter 14

Key Terms

normal biota
resident biota
transient biota
opportunist
changing biota
bifidus factor
symbiosis
commensalism
mutualism
microbial antagonism
parasitism
nonspecific surface defenses
mucous membrane
mucociliary system
lysozyme
adhesins
axillary
blepharitis

Study Tips

1. Have you completed a course on the anatomy and physiology of the human body? Review the structure and function of the skin and the various tracts of the body: respiratory, digestive, urinary, and reproductive. This knowledge will help you understand the environment of the normal biota described in this chapter.

2. A review of cell structure will also help you study this chapter. For example, how do some pathogens overcome the mucociliary system of the respiratory tract?

3. There are many applications of chemistry to the topic of microorganisms and human health. For example, how does a change in the pH of the small intestine (pH of 8 to 9) change the normal biota of this body region?

4. Compose a table to summarize the normal biota of the various regions of the body. List the body regions along the lefthand side of the table. Across from each region, describe the normal biota.

Normal Biota

A. Define each of the following and provide an example of a microorganism in the human body for each.

    1. resident biota

2. transient biota

3. opportunist

4. changing biota

B. Label each of the following body regions normally occupied by a normal biota population.

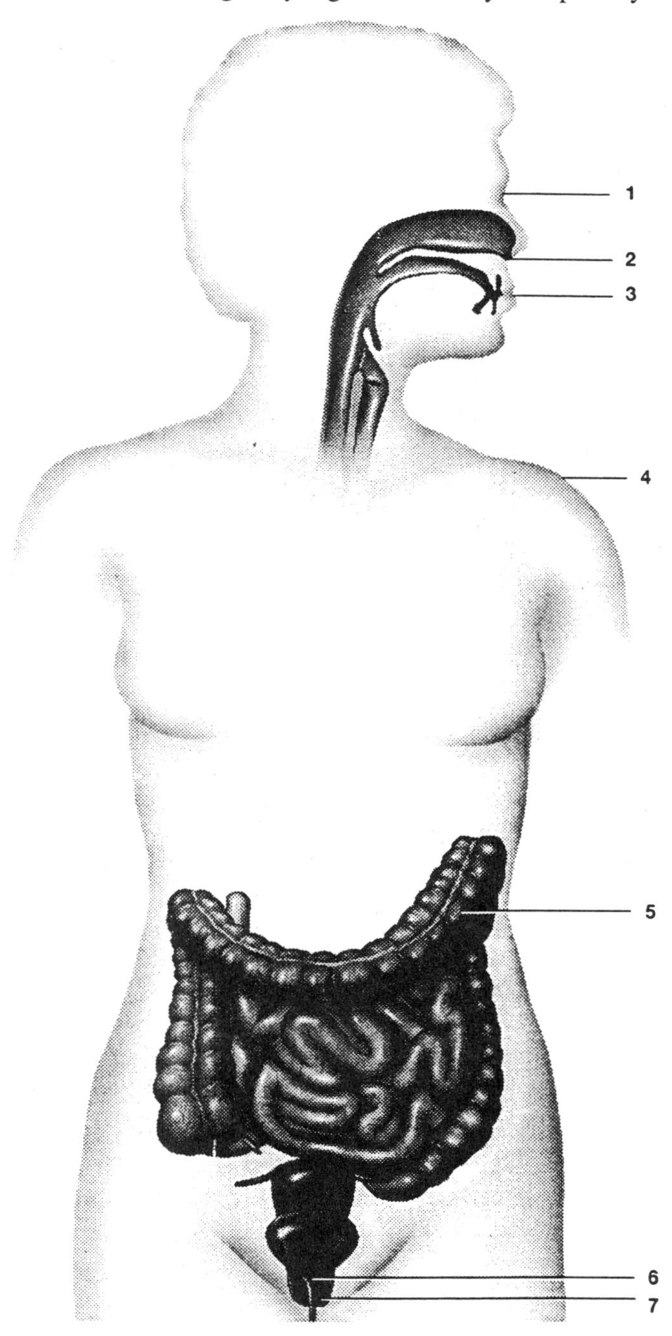

Chapter 14

Symbioses

A. Match each of the following descriptions to the correct kind of symbiosis.

1.	Both partners benefit.	A.	commensalism
2.	The host is harmed.	B.	mutualism
3.	One partner benefits and the other is neither helped nor harmed	C.	parasitism
4.	The majority of human normal biota are involved in this type of symbiosis.		
5.	This type is lacking in humans.		
6.	One member is a pathogen.		

Factors that Determine the Normal Biota

A. Label each of the following statements as true or false. If false, correct it.

1. Epithelial surfaces are a biochemical defense against microorganism invasion.
2. Compared to the skin, the conjunctiva is a thin membrane.
3. Surface skin cells are actively dividing.
4. The mucociliary system is an active defense mechanism of the digestive tract.
5. The rapid flow of urine discharges bacteria from the urethra.
6. Keratin is a component of the inner lining of the digestive tract.
7. The skin normally secretes HCl.
8. The stomach normally secretes fatty acids.
9. Gram-positive bacteria are particularly vulnerable to the effects of lysozyme.
10. Members of the normal biota of epithelial surfaces attach to the epithelium by their pili.

Sites of Normal Biota

A. Select the correct statement from each of the following pairs of statements.

1. A. The dryness of the skin limits microbial growth.
   B. The skin makes up less than five percent of the body weight.

2. A. The skin secretes fatty acids.
   B. The skin does not secete lysozyme.

3. A. The majority of the skin's normal biota are commensals.
   B. Fungi are not normal biota of the skin.

4. A. *Staphylococcus aureus* is found on everyone's skin.
   B. *Staphylococcus epidermidis* is found on everyone's skin.

5. A. Acne is caused by a virus.
   B. Aerobic diphtheroids grow new the surface of the skin.

6. A. Axillary normal biota live in the groin of a person.
   B. The conjunctiva is a mucous membrane.

7. A. Numerous microorganisms are inhaled daily.
   B. Cystic fibrosis patients produce a very thin mucus.

8. A. *Staphylococcus epidermidis* is found in the nasal cavity of most humans.
   B. *Moraxella catarrhalis* is a Gram-positive pathogen.

9. A. The mouth is an inhospitable environment for normal biota.
   B. The saliva contains lysozyme.

10. A. Staphylococci species are widespread in the oral cavity.
    B. Microbial populations in the mouth are unstable.

11. A. Meconium lacks microorganisms.
    B. The pH of the stomach is very acidic.

12. A. Bile salts destroy many microorganisms in the intestine.
    B. Peristalsis is the chemical digestion of food.

13. A. The large intestine has a large, diverse normal biota.
    B. *Clostridium* is normally absent in the large intestine.

14. A. The vagina does not sustain a normal biota population.
    B. The vagina has a low pH.

15. A. The upper urethra contains many microorganisms.
    B. The urethal lining has tightly joined epithelial cells.

Is the Normal Biota Helpful or Harmful?

A. 1. State one reason that normal biota can be helpful to the body.

2. State one reason that normal biota can be harmful to the body.

Discussion Questions

1. How do the concepts learned earlier in the text support the understanding of the major points in this chapter?

2. There are some species of normal biota that are the same in a given body region for all humans. However, there are also some differences in normal biota, region by region, among humans. What factors account for these differences?

Multiple Choice: Review

1. Select the human body structure that has a normal biota population.

   A. blood
   B. lower respiratory tract
   C. skin
   D. skeletal muscles

2. Opportunists are microorganisms that

   A. always cause a disease.
   B. cause disease when the proper condition arises.
   C. constantly produce beneficial effect in the host.
   D. never cause disease.

3. Which type of symbiosis is most common in the human body?

   A. amensalism
   B. commensalism
   C. mutualism
   D. parasitism

4. The mucociliary system of the human body is a ________ defense.

   A. biochemical
   B. mechanical
   C. structural
   D. transport

5. Adhesin helps normal flora attach to ____________ surfaces.

   A. connective tissue
   B. epithelial
   C. muscular
   D. nerve

6. Select the correct association.

   A. axillary/blood
   B. diphtheroids/Gram-negative rods
   C. *Propionibacterium*/species causes acne
   D. *Staphylococcus* species/strictly aerobic

7. The conjunctiva is a covering of the

   A. brain.
   B. eye.
   C. heart.
   D. kidney.

8. *Candida albicans* is a

   A. bacterium living in the oral cavity.
   B. bacterium living in the upper respiratory tract.
   C. yeast living in the oral cavity.
   D. yeast living in the upper respiratory tract.

9. Select the bacterium that is not found in substantial numbers in the bowel.

   A. *Bacteroides*
   B. *Bifidobacterium*
   C. *Escherichia*
   D. *Staphylococcus*

10. Microbially produced carcinogens are a main cause of cancer in humans.

    A. True
    B. False

Chapter 14

Answers

Normal Biota

A. 1. They are characteristic and permanent biota in a given body region. One example is *E. coli* living in the intestinal tract.

2. These biota cannot persist indefinitely in the body. Pathogenic strains of *Staphylococcus aureus* among hospital workers is one example.

3. These microorganisms normally do not inhabit the body. However, they will cause a disease in the body when the proper opportunity arises.

4. Some biota populations change in a body region over time. For example, intestinal biota change over the human lifespan.

B. 1. conjunctiva 2. nasal cavity and nasopharynx 3. mouth 4. skin 5. intestinal tract 6. urethra 7. vagina

Symbioses

A. 1. B 2. C 3. A 4. A 5. B 6. C

Factors that Determine the Normal Biota

A. 1. False; They are a structural defense.
2. True
3. False; They are dead and are replaced by the division of cells in the deeper layers of the skin.
4. False; It is a mechanism of the respiratory tract.
5. True
6. False; Keratin is a component of the outer layer of the skin.
7. False; The skin secretes fatty acids.
8. False; The stomach secretes HCl.
9. True
10. True

Sites of Normal Biota

A. 1. A 2. A 3. A 4. B 5. B 6. B 7. A 8. A 9. B 10. A 11. B 12. A 13. A 14. B 15. B

Is the Normal Biota Helpful or Harmful?

A. 1. A commensal can prevent a pathogen from establishing an environment in the body through microbial antagonism.

2. A commensal can become an opportunist and cause an infection in the body.

Discussion Questions

1. There are many examples for discussion. Differences in bacterial cell wall (Gram-positive versus Gram-negative) makeup accounts for the differences in lysozyme activity. pH differences throughout the body account for the optimal environments for the different normal biota.

2. Start by studying several characteristics: human diets, genetic differences, and variations in human physical development.

Multiple Choice: Review

1. C 2. B 3. B 4. B 5. B 6. C 7. B 8. C 9. D 10. B

Chapter 15
Microorganisms and Human Disease

The Seven Capabilities of a Pathogen
One: Maintaining a Reservoir
- Human Reservoirs
- Animal Reservoirs
- Environmental Reservoirs

Two: Getting and Entering a Host
- Portals of Entry
- Modes of Transmission
  - Respiratory Droplets
  - Fomites
  - Direct Body Contact
  - The Fecal-Oral Route
  - Arthropod Vectors
  - Parenteral Transmission

Three: Adhering to a Body Surface
Four: Invading the Body
Five: Evading the Body's Defenses
- Evading Phagocytosis
  - Capsules
  - Surface Proteins
- Evading the Immune System
  - Antigenic Variation
  - IgA Proteases
  - Serum Resistance
- Obtaining Iron

Six: Multiplying in the Host
- Toxins
  - Exotoxins
  - Endotoxins
- Other Toxic Proteins
- Damage Caused by Host Responses
- Viral Pathogenesis

Seven: Leaving the Body

Key Terms

virulence factor
transmission
portal of entry
fomite
sexually transmissible disease

vertical transmission
prenatal
perinatal
vector
airborne transmission
filamentous hemagglutinin
avirulent
noninvasive
endocytosis
phagocyte
intracellular pathogen
pathogenicity
antigenic variation
IgA protease
serum resistance
transferrin
lactoferrin
ferritin
siderophore
pathogenesis
toxin
hypersensitivity
exotoxin
endotoxin
adenyl cyclase system
antitoxin
toxoid
botulism
interleukin-1
cytolysin
autolysis

Study Tips

1. Interview a local ecologist or environmental biologist at your college. A brief visit with this person can provide you with additional insights about the concepts of this chapter.

2. Does your college have courses in the anatomy and physiology of the human body? Ask to arrange some study time in the lab for these courses. In the lab study the human torso model. Locate the different portals of entrance and exit described in this chapter.

3. Continue to answer the several kinds of review questions stated at the end of the chapter. Ask your instructor to provide answer keys to check the accuracy of your answers.

4. Continue to study vocabulary by writing out a list of key terms from the chapter. Define these terms in your own words.

The Seven Capabilities of a Pathogen

One: Maintaining a Reservoir

A. Define each of the following.

1. virulence factor -
2. incubatory carrier -
3. chronic carrier -
4. zoonosis -
5. animal reservoir -
6. environmental reservoir -

Two: Getting To and Entering a Host

A. Label each of the following statements as true or false. If false, correct it.

1. Transmission refers to a pathogen leaving its reservoir and entering the body of a host.
2. The phrase "infectious dose" refers to the number of microbial cells that must enter the host body to cause death in 50 percent of the test animals.
3. The phrase "lethal dose" refers to the number of microbial cells that must enter the host body to cause infection in 50 percent of the test animals.
4. *Bordetella pertussis* is transmitted from host to host by respiratory droplets.
5. Whooping cough is a communicable disease.
6. A fomite is a microorganism causing a disease.
7. The most effective approach to decrease the spread of colds is through frequent handwashing.
8. STDs are spread when the cutaneous membrane of an infected person contacts the cutaneous membrane of a noninfected person.

9. Herpes simplex I is caused by a bacterium.

10. A perinatal infection occurs across the placenta, before birth.

11. Some arthropod vectors are mechanical vectors.

12. Malaria is caused by a biological vector.

13. Airborne diseases must be spread by microorganisms constantly located in the air.

14. A species of *Mycobacterium* causes tuberculosis.

15. Parenteral transmission occurs by an infection through the digestive tract.

Three: Adhering to a Body Surface

Complete each of the following statements with the correct term.

1. Pathogens attach to body surfaces by structures called ________.

2. The cells or tissues that a pathogen attacks are the pathogen's tissue ________.

3. *Bordetella pertussis* cells attach to the _________ of the epithelial cells of the respiratory tract.

4. FHA is a wavy ________ and not a true pilus.

5. *Neisseria gonorrhoeae* binds to the genital, ________ pharyngeal, and __________ surfaces of the host body.

Four: Invading the Body

A. Short Answer

1. Name two noninvasive pathogens.

2. How do invasive pathogens enter cells of the host?

3. What is the function of phagocytes?

4. Describe the key to success of *Coxiella burnettii* to cause human infection.

5. What are invasins?

6. What is the unique property of intracellular pathogens?

Five: Evading the Body's Defenses

A. Select the correct statement from each of the following pairs.

1. A. *Streptococcus pneumoniae* is a capsule-dependent pathogen.
   B. Most bacteria attacking the nervous system lack a capsule.

2. A. *Streptotoccus pyrogens* produces an IgA protease.
   B. *Neisseria gonorrhoeae* produces Protein II.

3. A. Antigenic recognition allows a highly specific immune response.
   B. By antigenic variation a host changes its immune response.

4. A. IgA antibodies destroy IgA proteases.
   B. *Neisseria gonorrhoeae* makes an IgA protease.

5. A. Serum resistance acts against a complement system.
   B. Coalesce is an enzyme that destroys complement proteins.

6. A. Siderophores add iron to iron-transport proteins.
   B. Transferrin is an iron-transport protein.

Six: Multiplying in the Host

A. Matching - Each choice is used once.

1. endotoxin
2. hypersensitivity
3. interleukin-1
4. lipid A
5. meningococcemia
6. Ptx
7. tetanus toxin
8. toxoid

A. component used in a vaccine
B. condition caused by *Neisseria*
C. cytokine causing fever
D. disrupts adenyl cyclase system
E. exaggerated immune response
F. toxin activating complement
G. found in all Gram-negative bacteria
H. from *Clostridium*, attacks nervous system

B. Complete each of the following with the correct term or terms.

1. ________ ________ is an enzyme that potentiates the effect of the pertussis toxin in eukaryotic cells.

2. Cytolysins attack cell ________.

3. Collagen is a major structural component of ________ tissue.

4. Coagulase stimulates the conversion of ________ to ________.

5. Streptokinase is used to dissolve ________ ________.

6. Viruses destroy host cells from within by the process of ________.

7. During a ________ infection, a virus is dormant for years in a host cell.

8. The rabies virus produces ________ in infected nerve cells.

Seven: Leaving the Body

A. Name the portal of exit for each of the following.

1. respiratory pathogens

2. gastrointestinal pathogens

3. sexually transmitted pathogens

4. pathogens from arthropod vectors

Discussion Questions

1. Preview the upcoming chapters. How does the information in these chapters support the concepts presented in Chapter 15?

2. How does the information in earlier chapters support the concepts in Chapter 15?

Chapter 15

Multiple Choice: Review

1. A pathogen must have a reservoir in order to

   A. attach to a host.
   B. gain a good source of food.
   C. overcome the action of antibiotics.
   D. survive outside the human host.

2. Select the correct association.

   A. *Clostridium*/produces capsules
   B. fish/reservoir for Lyme disease
   C. *Salmonella*/causes chlolera
   D. *Streptococcus*/throat infection

3. *Bordetella pertussis* is spread between hosts by

   A. blood transfusions.
   B. contact between mucous membranes.
   C. contaminated food.
   D. respiratory droplets.

4. STD infections involve the contact of ________ membranes.

   A. cutaneous
   B. mucous
   C. serous
   D. synovial

5. A species of *Vibrio* is the causative agent of

   A. botulism.
   B. cholera.
   C. gonorrhea.
   D. malaria.

6. Biological vectors usually transmit infection by

   A. adhering to the mucous membranes.
   B. biting the host.
   C. entering through the gastrointestinal tract.
   D. entering through the respiratory tract.

7. FHA is a type of

    A. adhesin.
    B. antibody.
    C. antigen.
    D. white blood cell.

8. The species of *Plasmodium* enter the hosts'

    A. muscle cells.
    B. nerve cells.
    C. red blood cells.
    D. white blood cells.

9. Select the protein that does not transport iron in the human body.

    A. ferritin
    B. lactoferrin
    C. sideroferrin
    D. transferrin

10. Select the incorrect association.

    A. coagulase/breaks down fibrin
    B. hyaluronic acid/cements host cells together
    C. interleukin-1/stimulates fever production
    D. streptokinase/dissolves blood clots

Answers

The Seven Capabilities of a Pathogen

One: Maintaining a Reservoir

A. 1. The disease-causing capabilities of a microorganism, enabling it to cause an infection.

2. A person who is a human reservoir for a pathogen but who is apparently healthy. The person, however, does not harbor the pathogen chronically.

3. A person who is a human reservoir for a pathogen but who is apparently healthy and harbors the pathogen for years.

4. A human disease caused by a pathogen with an animal reservoir.

5. A animal that provides an environment for the survival of a pathogen.

6. A nonliving source that provides an environment for the survival of a pathogen.

Two: Getting To and Entering a Host

A. 1. True
2. False; The infectious dose is the number of cells needed to cause *infection* in 50% of the test animals.
3. False; The lethal dose is the number of cells needed to cause *death* in 50% of the test animals.
4. True
5. True
6. False; A fomite is an inanimate object serving as a vehicle for transmitting a pathogen.
7. True
8. False; STD transmission involves the contact of mucous membranes.
9. False; Herpes simplex I is caused by a virus.
10. False; A perinatal infection occurs during the passage through the birth canal or shortly after birth.
11. True
12. True
13. False; Airborne microbes, causing infections, may settle on dust particles and become airborne again.
14. True
15. False; A parenteral infection occurs by transmission through a break in the skin or mucous membranes.

Three: Adhering to a Body Surface

A. 1. adhesins 2. trophism 3. cilia 4. filament 5. rectal, conjunctival

Four: Invading the Body

A. 1. *Bordetella pertussis* and *Streptococcus pneumoniae*
2. endocytosis
3. They engulf and destroy invading microbes.
4. It can overcome phagocytosis by the host cells they invade.
5. Invasins are proteins on a pathogen's surface that stimulate phagocytosis attacking a host cell.
6. Intracellular pathogens stay inside host cells, multiplying there and not spreading.

Five: Evading the Body's Defenses

1. A 2. B 3. A 4. B 5. A 6. B

Six: Multiplying in the Host

A. 1. G 2. E 3. C 4. F 5. B 6. D 7. H 8. A

B. 1. adenyl cyclase 2. membranes 3. connective 4. fibrinogen to fibrin 5. blood clots 6. autolysis 7. latent 8. Negri bodies

Seven: Leaving the Body

A. 1. Respiratory pathogens leave through the nose by respiratory secretions.

2. Gastrointestinal pathogens leave by the anus.

3. STD pathogens leave across the mucous membrane surfaces of the genital tract.

4. Pathogens from arthropod vectors leave via a drop of blood.

Discussion Questions

1. The upcoming chapters describe the mechanisms of immunity to fight off the diseases caused by microbes.

2. There are many examples. Topics include viral structure, bacterial morphology (e.g., pili), and Gram-staining.

Multiple Choice: Review

1. D 2. D 3. D 4. B 5. B 6. B 7. A 8. C 9. C 10. A

Chapter 16
Defending the Body's Interior: Nonspecific Defenses

- Inflammation
    - Inflammatory Stimuli
    - Inflammatory Mediators
        - Bacterial By-Products
        - Complement Fragments
        - Kinins
        - Histamine
        - Prostaglandins and Leukotrienes
    - Effects of Inflammatory Mediators
        - Altering Capillaries
        - Attracting and Stimulating Phagocytes
    - Acute Inflammation
    - Inflammatory Repair
- Leukocytes
    - The Phagocytic Family
        - Polymorphonuclear Leukocytes
            - Neutrophils
            - Eosinophils
            - Basophils (and Mast Cells)
        - Mononuclear Leukocytes
            - Monocytes
            - Macrophages
    - Phagocytosis
- Complement
    - Complement Cascades
    - The Classical Pathway
    - The Alternate Pathway
    - The Terminal Pathway
- Interferon

Key Terms

inflammation
phagocytosis
complement
interferon
leukocyte
phagocytosis
inflammatory mediator
vasodilation
permeability

immune defense
kinase
mast cell
prostaglandin
leukotriene
chemotactic
margination
diapedesis
neutrophil
eosinophil
basophil
monocyte
lymphocyte
degranulation
allergy
macrophage
opsonin

Study Tips

1. A study of physiology will you help to understand the concepts in this chapter. Review the functions of macrophages and mast cells as well as hematology.

2. Can you reserve some time in your biology lab to study the human skeleton? Use the skeleton as a reference to study the various sites of blood cell production.

3. Also, use the lab as an opportunity to study the following cells under the microscope: mast cell, macrophage, neutrophil, eosinophil, basophil, lymphocyte, and monocyte.

4. Do you under the mathematical concepts presented in this chapter? Ask your instructor for some examples or make up your own study questions. For example, if the total white blood cell count in a patient is 8000 cells per cubic mm, what is the most frequent white blood cell type in this total? If its number is 5000 of the total, how is its percentage expressed in a differential white blood cell count?

The Body's Three Lines of Defense Against Infection

A. Short Answer

1. Name the three nonspecific surface defenses.

2. Name the four nonspecific interior defenses.

Chapter 16

Inflammation

A. Select the correct statements from among the following.

1. Inflammatory mediators act on capillaries to produce vasodilation.
2. The vasodilation of capillaries decreases their permeability.
3. F-met-leu-phe is a short segment of bonded monosaccharides.
4. Bradykinin is a vasodilator.
5. Histamine is released by erythrocytes.
6. Prostaglandins act as inflammatory mediators.
7. Some leukotrines attract leukocytes to sites of inflammation.
8. Erythema is an indication of inflammation of the skin.
9. Phagocytes adhere to the walls of capillaries by the process of margination.
10. Diapedesis is the breakdown of white blood cells during phagocytic attack.
11. Complement can kill human cells.
12. As scavengers, macrophages serve in the first step of inflammatory repair.

Leukocytes

A. Completion - Complete each of the following with the correct term or terms.

1. The two families of leukocytes are the _________ family and the ________ family.
2. A __________ blood count tallies the total number of white blood cells in a person's blood.
3. A ___________ white blood cell count tallies the percentages of the various kinds of leukocytes in a person's blood.
4. Segs are ____________ neutrophils with segmented ________.
5. Bands are ________ neutrophils with incompletely segmented ________.
6. Hematopoietic stem cells are __________ cells.

7. When stem cells differentiate, they __________ into different kinds of white blood cells.

B. Matching - Match each description to the correct kind of white blood cell. Each description matches to only one letter or description. However, a letter can be used more than once throughout the match.

1. 250 per cubic mm
2. 40 per cubic mm
3. some are wandering
4. some are fixed
5. 5000 per cubic mm
6. It secretes lactoferrin.
7. Its cytoplasmic granules stain blue.
8. It acts against parasitic worms.
9. rare cells that play a major role in allergies
10. 400 per cubic mm

A. basophil
B. eosinophil
C. macrophage
D. monocyte
E. neutrophil

C. Briefly describe each of the steps of phagocytosis.

1. activation -
2. chemotaxis -
3. recognition and adherence -
4. ingestion -
5. killing and digestion -
6. expulsion -

Complement

A. Label each of the following as describing the classical pathway, alternate pathway, or terminal pathway.

1. Activation does not require an antigen-antibody complex.

2. Activation requires an antigen-antibody complex.

3. It is a part of the body's third line of defense.

4. It is part of the body's nonspecific secondary line of defense.

5. Factor B, factor D, properdin, and C3 come together on the surface of a bacterial cell.

6. C9 molecules associate to become a membrane attack complex.

7. Gram-negative bacteria become vulnerable to the cytolysis by this pathway.

Interferon

A. Completion - Complete each of the following with the correct term.

1. Alpha interferon is produced by a type of lymphocyte _______ cell.

2. Gamma interferon is produced by a type of lymphocyte ________ cell.

3. Beta interferon is produced by ________.

4. AVPs are __________ that interfere with viral protein synthesis.

5. Interferons are virus-________, meaning that the same interferon is active against many different viruses.

Discussion Questions

1. Often the extracellular fluid of the human body is labeled the internal environment of the body. Explain this reference.

2. The term "homeostasis" refers to maintaining relatively constant conditions in the body's internal environment. How do the three line of defense against infection contribute to body homeostasis?

Multiple Choice: Review

1. Inflammatory mediators usually cause the

   A. vasoconstriction of blood vessels.
   B. vasodilation of blood vessels.

2. Complement components act primarily on

   A. blood vessels.
   B. brain cell receptors.

3. Histamine is a

   A. vasoconstrictor.
   B. vasodilator.

4. During erythema, there is a(n)__________ bloodflow to the skin.

   A. decreased
   B. increased

5. Phagocytes migrate and adhere to a capillary wall by the process of

   A. diapedesis.
   B. margination.

6. Select the agranular leukocyte.

   A. basophil
   B. eosinophil
   C. neutrophil
   D. monocyte

7. Select the most abundant white blood cell in a differential white blood cell count.

   A. basophil
   B. eosinophil
   C. neutrophil
   D. monocyte

8. Select the first step of phagocytosis.

   A. activation
   B. chemotaxis
   C. digestion
   D. recognition

9. Which complement pathway does not require an antigen-antibody complex?

   A. alternate
   B. classical

10. Interferons are

    A. virus-nonspecific.
    B. virus-specific.

Answers

The Body's Three Lines of Defense

A. 1. They are structural, mechanical, and biochemical.

   2. They are inflammation, phagocytosis, complement, and interferon.

Inflammation

A. 1, 4, 6, 7, 8, 9, 11, 12

Leukocytes

A. 1. phagocytic, lymphoid 2. complete 3. differential 4. mature, nuclei 5. immature, nuclei 6. blood-forming 7. specialize

B. 1. B 2. A 3. C 4. C 5. E 6. E 7. A 8. B 9. A 10. D

C. 1. A phagocyte produces peptidoglycan. This allows it to recognize foreign objects.

   2. Phagocytes are attracted to foreign objects.

   3. The phagocyte attaches to the bacterium. It is affected by bacterial capsules and opsonins.

   4. By pseudopods, the phagocyte engulfs the bacterium.

   5. Lysozyme activity breaks down the ingested bacterium.

   6. The phagocyte expels indigestible debris.

Complement

A. 1. alternate 2. classical 3. classical 4. Alternate 5. alternate 6. terminal 7. terminal

Interferon

A. 1. T 2. T 3. fibroblasts 4. enzymes 5. nonspecific

Discussion Questions

1. Although outside the cells, the extracellular fluid (ECF) is within the body. About two-thirds of human body fluid is intracellular. The other one-third, the ECF, consists of the blood plasma and tissue fluid.

2. The body's three lines of defense keep the blood plasma and tissue fluid free of pathogens. Therefore, these lines contribute to maintenance of a normal biota throughout the human body while ridding the body of abnormal, unwanted microbes.

Multiple Choice: Review

1. B 2. A 3. B 4. B 5. B 6. D 7. C 8. A 9. A 10. A

Chapter 17
The Immune System - Specific Defenses of the Body's Interior

Study Tips

1. Continue to answer the review questions, correlation questions, and essay questions at the end of the chapters. Discuss the answers with your classmates. Ask your instructor to post an answer key to check for the accuracy of your responses.

2. Is there a human torso model in your biology lab? Refer to it to locate the different organs in the body with lymphatic tissue.

3. If there is a long bone in your biology lab, study it to locate the sites of blood cell formation from stem cells in the marrow cavity.

4. Use a molecular model kit to build the basic makeup of an antibody molecule. If such a kit is unavailable, sketch the basic makeup from memory.

Cells and Organs of the Immune System

A. Four of the following eight statements are true. Select the true statements.

1. Among the three lines of defense, the immune system is the most powerful line.
2. The immune system is effective against viruses and not nearly as effective against bacteria.
3. Among the different kinds of white blood cells, lymphocytes are comparatively large cells.
4. B cells respond to antigens by producing antibodies.
5. T cells are capable of producing lymphokines.
6. The bone marrow and the thymus are secondary lymphoid organs.
7. The lymph nodes are primary lymphoid organs.
8. Lymphocytes circulate through the lymphatic system and the blood stream.

An Immune Reaction: An Overview

A. Completion - Complete each of the following with the correct term.

1. Almost all __________ molecules are strong antigens.
2. If a foreign nucleic acid or ___________ molecule enters the human body, it will probably not stimulate an immune response.

3. Immune ________ is the ability not to react to one's own body molecules.

4. A lymphocyte will not recognize an antigen if it loses its antigen ________.

5. Another name for an antigenic determinant is an ________.

6. A lymphocyte receptor will not recognize its complement antigenic determinant if its _________ changes.

B Cells

A. Label each of the following statements as true or false. If false, correct it.

1. During differentiation in the bone marrow, B cells become immunocompetent.

2. There are many more human genes than antibodies.

3. During differentiation individual B lymphocytes make their own unique antibodies through genetic recombination.

4. B-cell activation triggers B-cell recognition.

5. Clonal selection produces a group of identical B cells.

6. Lymphokines are substances that kill lymphocytes.

7. Macrophages secrete interleukin-1.

8. B-cell activation is a one-step process.

9. When an infection is over, plasma cells continue to circulate in the body.

10. A secondary immune response takes weeks to develop.

11. A secondary immune response is initiated by memory B cells.

12. The fit between an antigen and its specific antibody is somewhat variable.

B. Select the correct statement from each of the following pairs.

1. A. Members of five antibody groups share a basic X structure.
   B. Among the five antibody groups, component monomers vary.

2. A. Antibody monomers consist of four polypeptide chains.
   B. Polypeptide monomers are united by ionic bonds.

3. A. The ends of the antibody molecule are variable.
   B. The stem of the Y-shaped antibody bonds to the antigen.

4. A. An antibody monomer can bind to two epitopes.
   B. An antibody with more than one monomer has a valence of 1.

C. Match each class of antibody to its correct description.

1. IgA	A. They make up 75% of the antibodies.
2. IgD	B. They produce the secretory component.
3. IgE	C. They are sometimes called the early antibody.
4. IgG	D. They are poorly understood.
5. IgM	E. They can cause harmful allergic reactions.

T Cells

A. Four of the following eight statements are true. Select the correct statements.

1. T cells are able to control intracellular infections.
2. T cells cannot produce lymphokines.
3. Dendritic cells are antigen-presenting cells.
4. MHC are specialized genes making antibody molecules.
5. T lymphocytes are in a resting state until their antigenic receptors encounter and bind to a matching antigen fragment.
6. Human cells cannot signal that they have been infected by a virus.
7. Perforins secreted by TC cells mark infected cells for antibody activation.
8. Human cells can signal that they have been infected by bacteria.

Non-B Non-T Cells

A. Completion - Complete each of the following with the correct term.

1. Non-B non-T cells function without recognizing a(n) ________.

2. Non-B non-T lymphocytes are sometimes called ________ lymphocytes.

3. NK cells target and ___________ human cells.

4. NK cells can respond against _________ organs.

Immune Tolerance

A. Completion

Complete the following paragraph with the correct terms.

Differentiating __1__ cells in the bone marrow and __2__ cells in the thymus have not fully specialized. They undergo __3__, which is programmed cell death. By reacting with self antigens, these cells are __4__ during development. This process is called clonal __5__.

Types of Immunity

A. Short Answer

1. Name the four basic kinds of immunity.

Discussion Questions

1. How do you think autoimmunity develops against a structure of the human body?

2. How is the concept of cell specialization demonstrated in this chapter?

Multiple Choice: Review

1. Humoral immunity means that the antibodies are

A. constantly changing.
B. destroyed in the serum.
C. dissolved in fluid.
D. reacting with antigens.

2. Select the primary lymphoid organs.

   A. bone marrow and thymus
   B. bone marrow and lymph nodes
   C. lymph nodes and spleen
   D. spleen and thymus

3. An epitope is a

   A. differentiated B cell.
   B. differentiated T cell.
   C. small part of an antibody.
   D. small part of an antigen.

4. Which cell becomes immunocompetent?

   A. T cell
   B. B cell

5. By which process do B cells proliferate?

   A. clonal selection
   B. immune tolerance

6. Which cell type can differentiate into a plasma cell?

   A. T cell
   B. B cell

7. Which immune response generates more antibodies?

   A. primary
   B. secondary

8. For an antibody molecule, the variable regions are at the

   A. end of an X shaped molecule.
   B. end of a Y shaped molecule.
   C. middle of an Y shaped molecule.
   D. middle of an X shaped molecule.

9. Which class of antibodies protect mucosal surfaces?

   A. IgA
   B. IgB
   C. IgE
   D. IgM

10. Which type of immunity is stimulated by vaccines?

   A. artificially acquired active
   B. artificially acquired passive
   C. naturally acquired active
   D. naturally acquired passive

Answers

Cells and Organs of the Immune System

A. 1, 4, 5, 8

An Immune Reaction: An Overview

1. protein 2. lipid 3. tolerance 4. receptor 5. epitope 6. shape

B Cells

A. 1. True
   2. False; There are 100,000 human genes and 100 million human antibodies.
   3. True
   4. False; B cell recognition triggers B cell activation.
   5. True
   6. False; Lymphokines are molecules produced by lymphocytes and are necessary for the immune response.
   7. True
   8. False; B cell activation is an incremental process.
   9. False; Plasma cells die when the infection is over, but memory cells circulate.
   10. False; The secondary immune response requires only a few days.
   11. True
   12. True

B. 1. B 2. A 3. A 4. A

C. 1. B 2. D 3. E 4. A 5. C

T Cells

A. 1, 3, 5, 8

Non-B Non-T Cells

A. 1. antigen 2. granular 3. lyse 4. transplanted

Immune Tolerance

A. 1. B 2. T 3. apoptosis 4. eliminated 5. deletion

Types of Immunity

A. 1. naturally acquired active immunity, artificially acquired active immunity, naturally acquired passive immunity, artificially acquired passive immunity

Discussion Questions

1. A particular line of immune cells is not eliminated through clonal deletion early in development. If this line of cells survives, it can carry out an immune response against a specific protein in a structure such as the cornea of the eye or the kidney.

2. The variety of T cells offers one example. They are derived from an identical genetic source. However, their varieties result from differential gene expression. The wide variety of antibody production through genetic recombination is another example.

Multiple Choice: Review

1. C 2. A 3. D 4. B 5. A 6. B 7. B 8. B 9. A 10. A

Chapter 18
Immunologic Disorders

Immune System Malfunctions
Hypersensitivity
  Type I: Anaphylactic Hypersensitivity (Allergy)
    Immunologic Basis of Type I Hypersensitivity
    Allergies
  Type II: Cytotoxic Hypersensitivity
    Immunologic Basis of Type II Hypersensitivity
    Autoimmune Diseases
    Transfusion Reactions
    Hemolytic Disease of the Newborn
  Type III: Immune-Complex Hypersensitivity
    Immunologic Basis of Type III Hypersensitivity
    Systemic Lupus Erythematosus
    Other Examples of Type III Hypersensitivity
  Type IV: Cell-Mediated (Delayed) Hypersensitivity
    Immunologic Basis of Type IV Hypersensitivity
    Contact Hypersensitivity
    Tuberculin Skin Test
    Granulomatous Reaction
Organ Transplantation
  Tissue Typing and Matching for Transplantation
  Immunosuppression
Immunodeficiencies
  Congenital Immunodeficiencies
  Acquired Immunodeficiencies
Cancer and the Immune System
Summary

Key Terms

hypersensitivity
immunodeficiency
anaphylaxis
sign
symptom
allergic rhinitis
asthma
Grave's disease
Goodpasture's syndrome
major crossmatch
minor crossmatch

hemolytic disease
Rh-immunized
systemic lupus erythematosus
granulomatous reaction
contact hypersensitivity
granuloma
autograft
isograft
allograft
xenograft
immunologically privileged site
human leukocyte antigen complex
tissue type
immunosuppression
X-linked agammaglobulinemia
immune surveillance theory
cancer immunotherapy

Study Tips

1. Interview a local health professional on your campus or nearby community. This communication can provide you with additional insights to understand the concepts in this chapter.

2. Arrange to study transverse sections of blood vessels and bronchi/bronchioles in lab. Knowledge of their histology (tissue makeup) will help you to understand some of the reactions described in the chapters on immunity. For example, when the smooth muscle in the walls of these structures contracts, these structures constrict. Relaxation of the muscle results in dilation.

3. Use of an artificial blood typing kit in lab can help you to visualize the antigen-antibody reactions of the ABO and Rh blood groups.

4. Continue to refer to the human torso model in lab. Locate structures such as the thymus gland, spleen, and lymph nodes.

Immune System Malfunctions

Hypersensitivity

A. Select the correct statements about anaphylactic hypersensitivity.

   1. Its more common name is allergy.

   2. Sensitization by an allergen leads to a harmful allergic reaction with future exposure to an allergen.

3. The attachment of IgA antibodies to basophils is a major part of this response.
4. Histamine, an inflammatory mediator, leads to massive vasoconstriction of blood vessels.
5. Leukotrines are formed by degranulation and act slowly.
6. Inflammatory mediators produce the signs and symptoms.
7. Bronchial smooth muscles relax during this response.
8. Edema of the skin develops.
9. Fluid is lost rapidly from the circulation.
10. As an effective treatment, the administration of epinephrine constricts blood vessels and relaxes bronchial muscle.
11. This type of reaction is very rare.
12. Inhaled allergens can cause allergic rhinitis.
13. Hay fever is primarily mediated by histamine.
14. Inhaled allergens that affect the upper respiratory tract cause asthma.
15. Bronchi narrow during the response in asthma.
16. Histamine is mainly responsible for asthma.
17. Respiratory allergies are detected by skin testing.
18. Desensitization is effective at inhibiting respiratory allergens.
19. Ingested allergens are relatively harmless.
20. Penicillin is a hapten that can become an allergen.

B. Select the correct statements about cytotoxic hypersensitivity.

1. IgM and IgG antibodies bind abnormally to body cells.
2. This response may activate the complement pathway.
3. All autoimmune diseases are caused by type II sensitivity.
4. Grave's disease develops when antibodies bind to cells that produce human growth hormone.

5. In Goodpasture's syndrome, body cells are killed.

6. The ABO blood groups are name by the antibodies present.

7. The ABO antibodies are of the type IgM class.

8. Agglutination is the lysing of erythrocytes.

9. A person with type B blood can receive type O blood without agglutination.

10. A person with type A blood can donate that blood to a person with type AB blood without agglutination.

11. Rh negative blood lacks the Rh factor.

12. Production of anti-Rh antibodies requires exposure to Rh-positive blood.

13. The hemolytic disease of the newborn involves an Rh negative baby.

14. The hemolytic disease of the newborn involves an Rh positive mother.

15. The hemolytic disease of the newborn can be prevented by preventing the Rh-immunization of the Rh negative mother.

C. Select the correct statements about Immune-complex hypersensitivity.

1. IgG and IgM antibodies react with a person's body cells.

2. Circulating, soluble antigen-antibody complexes initiate this kind of reaction.

3. Complement is activated in capillaries.

4. SLE can be effectively treated.

5. Rheumatoid arthritis is an autoimmune disease.

D. Select the correct statements about cell-mediated hypersensitivity.

1. It is not triggered by antibodies.

2. This response occurs within a few minutes after exposure.

3. The granulomatous reaction causes tissue damage.

4. Poison ivy occurs by contact hypersensitivity.

5. Uroshiol is the sensitizing hapten for causing poison ivy.

6. TD cells release lymphokines leading to granuloma formation.

Organ Transplantation

A. Matching - Each transplant description is used once.

1. allograft	A. from nonhuman primate to human
2. autograft	B. between genetically identical individuals
3. isograft	C. involving only one organism
4. xenograft	D. between genetically different humans

B. Label each of the following statements as true or false. If false, correct it.

1. Immunologically privileged sites are not subject to tissue rejection.

2. MHC antigens are the primary antigens considered in tissue matching for possible transplantations.

3. MHC antigens are encoded by the HLA complex.

4. A person's tissue type is determined by mixing the person's complement and antibodies against antigen-bearing lymphocytes.

5. The HLA system has only four antigenic forms.

6. Cyclosporine interferes specifically with T-cell function.

Immunodeficiencies

A. 1. SCID involves disorders that disable both _______ cell and ________ cell immunity.

2. Patients with X-linked agammaglobulinemia have almost normal ____________ mediated immunity.

3. DiGeorge's syndrome affects only ___________ cell function.

4. In granulomatous, ____________ are cells that lack the ability to produce a respiratory burst.

5. Lymphoma is cancer of the ________ __________.

Cancer and the Immune System

A. Short Answer

1. How many TC cells, NK cells, and macrophages play a role in controlling cancer?

2. What is the immune surveillance theory?

3. What is the immune escape theory?

4. Why may the immune system serve as a safe way to treat cancer when compared to other techniques?

Discussion Questions

1. The thymus gland tends to atrophy in the human adult. Can you explain the basis for this? How do you think this change affects the immune system in the adult?

2. What do you think a complete cross-matching of the blood involves? Are ABO and Rh the only blood groups involved?

Multiple Choice: Review

1. Hypersensitivity is a __________ response.

   A. rapid
   B. slow

2. During an allergic reaction, bronchi

   A. constrict.
   B. dilate.

3. The action of inflammatory mediators leads to

   A. blood fluid gain.
   B. blood fluid loss.

4. During asthma inhaled allergens affect the ________ respiratory tract.

   A. upper
   B. lower

5. The autoantibodies of Grave's disease are

   A. IgA.
   B. IgD.
   C. IgE.
   D. IgG.

6. Bloodtype A can donate to bloodtypes ____________ without causing agglutination.

   A. A and O
   B. A and AB
   C. B and O
   D. B and AB

7. A person who is Rh positive usually __1__ the Rh factor and __2__ the anti-Rh antibody.

   A. 1 - possesses, 2 - possesses
   B. 1 - possesses, 2 - lacks
   C. 1 - lacks, 2 - possesses
   D. 1 - lacks, 2 - lacks

8. Rheumatoid arthritis is a type ____________ hypersensitivity.

   A. I
   B. II
   C. III
   D. IV

9. Which type of organ transplant only involves one organism's body?

   A. allograft
   B. autograft
   C. isograft
   D. xenograft

10. Cyclosporine is a product of a(n)

    A. alga.
    B. bacterium.
    C. fungus.
    D. protozoan.

Answers

Immune System Malfunctions

Hypersensitivity

A. 1, 2, 5, 6, 8, 9, 10, 12, 13, 15, 17, 20
B. 1, 2, 5, 7, 9, 10, 11, 12, 15
C. 2, 3, 5
D. 1, 3, 4, 5, 6

Organ Transplantation

A. 1. D 2. C 3. B 4. A

B.
1. True
2. True
3. True
4. False; A person's antigen-bearing lymphocytes are mixed with complement and antibodies against different HLA antigens.
5. False; The HLA system has more than 10,000 antigenic forms.
6. False, Cyclosporin interferes with T-cell function.

Immunodeficiencies

A.
1. B and T
2. cell
3. T
4. neutrophils
5. lymph nodes

Cancer and the Immune System

A.
1. They destroy antigen-marked cells that are transformed in the development of cancer.

2. The transformation of cancer cells occurs often, but the immune system usually eliminates these cancer cell.

3. Cancer cells do not stimulate an immune response or they sometime escape the immune response, leading to the development of the disease.

4. Many cancer cells are nonspecific and harm other body cells in addition to cancer cells. The immune system can be more selective.

Chapter 18

Discussion Questions

1. The thymus gland is located in the mediastinum of the human body. With its atrophy, other body sites must process T cells.

2. There are dozens of antigen-antibody systems in the human body that must be checked for cross-matching. For example, the MN system must be checked in addition to ABO and Rh.

Multiple Choice: Review

1. A 2. A 3. B 4. B 5. D 6. B 7. B 8. C 9. B 10. C

Chapter 19
Diagnostic Immunology

- Diagnostic Immunology
- Detecting Antigen-Antibody Reactions
  - Precipitation Reactions
    - Simple Immunodiffusion Tests
    - Tests that First Separate Antigens
      - Immunoelectrophoresis
      - Western Blots
  - Agglutination Reactions
  - Complement Fixation Reactions
- Immunoassays
- Fluorescent Antibodies
- Summary

Key Terms

diagnostic immunology
monoclonal antibody
serology
serum
test reagent
precipitation reactions
equivalence zone
double diffusion method
radial diffusion
electrophoresis
direct agglutination
indirect agglutination
titer
indicator system
radioimmunoassay
immunofluorescence assay
solid phase immunosorbent assay
enzyme-linked immunosorbent assay
fluorescent antibody
fluorescein isothiocyanate

Study Tips

1. Visit a local faculty member or research scientist conducting studies in the fields described in this chapter. If you cannot locate this person, ask your instructor for advice in your search.

2. Are any of your lab activities offering opportunities to study the concepts of diagnostic immunology? Ask your instructor for examples.

3. Use molecular models to demonstrate the various bonding patterns between antigens and antibodies.

Diagnostic Immunology

A. Completion - Complete each of the following with the correct term.

1. Diagnostic immunology is a field that uses antigen -antibody reactions to diagnose ________.

2. The pregnancy test uses ____________ antibodies.

3. A blood sample is about 45 percent cellular, by volume. The percent that is serum is about ________ percent.

4. In a diagnostic immunology test, __________ in a test reagent are used to detect antigens.

5. A test is conducted to learn if a person is infected by hepatitis A. Diagnosis involves detecting antibodies that a person makes against the ___________ molecules of a virus.

Detecting Antigen-Antibody Reactions

A. Select the correct statements among the following.

1. All serologic tests involve a reaction between an antigen and an antibody.

2. After a chemical reaction, the contents in a test tube changes from cloudy to clear. This is an indication of an antigen-antibody reaction.

3. A certain antigen and antibody each has one binding site. This will allow them to react and build a molecular lattice.

4. For an immunodiffusion test, an antibody in a well spreads through the surrounding gel. Its concentration in the gel increases with increasing distance from the well.

5. In a simple immunodiffusion test, a precipitate between diffusing antigen and antibody will form in the zone of equivalence.

6. In a double diffusion test, two antigens form a continuous line on two sides of an antibody well. This result means that the antigens are identical.

7. In a radial diffusion test, an increasing ring in the size of a precipitate means increasing antigen concentration.

8. Immunoelectrophoresis is used to separate antigens in a mixture.

9. A western blot test is less sensitive that immunoelectrophoresis.

10. In the agglutination reaction to detect human blood type, antigen A reacts with antibody A and forms a clump of cells.

11. In the agglutination reaction to detect human blood type, antigen B is on the surface of human erythrocytes. It is detected by reacting with antibody A.

12. The Coombs test is a sensitive test.

13. Agglutination occurs when cells do not clump together.

14. A titer is the lowest dilution of a test serum that causes agglutination.

15. In a complement fixation test, lysis of erythrocytes occurs. This indicates that anti-erythrocyte antibody was present.

Immunoassays

A. Label each of the following statements about immunoassays as true or false. If false, correct it.

1. They are extremely sensitive.

2. Antigens cannot be detected.

3. Antibodies can be detected.

4. To be performed successfully, an antigen-antibody product does not need to be separated from a tagged reactant.

5. ELISA are becoming the dominant immunoassays.

Fluorescent Antibodies

A. Short Answer

1. State a major use in the biology lab for fluorescent antibodies.

Chapter 19

Discussion Questions

1. What are some of the practical uses of diagnostic immunology?

2. What is the difference between the serum and plasma of the blood?

Multiple Choice: Review

1. In diagnostic immunology

   A. only an antigen can be used to detect an antibody.
   B. only an antibody can be used to detect an antigen.
   C. antigens and antibodies can be used to detect each other.
   D. antigens and antibodies are not used.

2. Serologic tests are

   A. quantitative only.
   B. qualitative only.
   C. quantitative and qualitative.
   D. neither quantitative nor qualitative.

3. For a precipitation reaction between an antigen and antibody, each molecule must have at least __________ binding sites.

   A. two
   B. four
   C. six
   D. eight

4. For an antigen-antibody reaction, cloudiness in a test tube incidates

   A. a precipitate and an antigen-antibody reaction.
   B. a precipitate without an antigen-antibody reaction.
   C. no precipitate and an antigen-antibody reaction.
   D. no precipitate without an antigen-antibody reaction.

5. In an electrophoretic field, the anode is the negative field. A positive molecule will migrate

   A. away from the anode.
   B. toward the anode.

6. Hemagglutination is a(n) __________ agglutination reaction.

   A. direct
   B. indirect

7. The ABO testing for human blood type is an ________ test.

   A. qualitative
   B. quantitative

8. A persons blood does not produce any agglutination in the ABO blood typing test. The person's ABO blood type is

   A. A
   B. B
   C. O
   D. AB

9. The complement fixation test is still very important for detecting a particular

   A. alga.
   B. bacterium.
   C. protozoan.
   D. virus.

10. ELISA are __________ sensitive than most immunoassays.

   A. less
   B. more

Answers

Diagnostic Immunology

A. 1. diseases 2. monoclonal 3. fifty-five 4. antibodies 5. antigen

Detecting Antigen-Antibody Reactions

A. 1, 5, 6, 7, 8, 10, 12

Immunoassays

A. 1. True

2. False; Antigens and antibodies can be detected.
3. True
4. False; They antigen and antibody must be separated.
5. True

Fluorescent Antibodies

A. 1. They are used to visualize specific antigens in tissues or on the surface of microorganisms.

Discussion Questions

1. There is a diagnostic value to this branch of science. As one example, start with diagnosing hepatitis A by identifying the antibodies that the human body makes against the virus causing this disease.

2. Plasma is the complete, liquid part of the blood. Serum is similar to plasma, except that it does not have the components used by the blood for blood coagulation.

Multiple Choice: Review

1. C 2. C 3. A 4. A 5. B 6. A 7. A 8. C 9. D 10. B

Chapter 20
Preventing Disease: Epidemiology and Public Health

Preventing Disease
Epidemiology
- The Methods of Epidemiology
  - Sources of Information
  - Uses of Statistics
- Types of Epidemiological Studies
  - Descriptive Epidemiology
  - Surveillance Epidemiology
  - Field Epidemiology
  - Hospital Epidemiology
    - Nosocomial Infections
    - Types of Nosocomial Infections
    - Hospital Epidemiologists and Infection Control

Public Health
- Public Health Organizations
- Diminishing Reservoirs and Interfering with Transmission
  - Clean Water
  - Clean Food
  - Personal Cleanliness
  - Insect Control
  - Prevention of Sexually Transmissible Disease
  - Prevention of Respiratory Diseases

Immunization
- Active Immunization
  - Types of Vaccines
    - Attenuated Vaccines
    - Inactivated Vaccines
    - Acellular Vaccines
    - DNA Vaccines
  - Active Immunization and Public Health Policy
- Passive Immunization

Summary

Key Terms

epidemic
pandemic
endemic
sporadic
notifiable disease
incidence rate

Chapter 20

endoscopy
bronchoscopy
laparoscopy
urinary catheterization
prophylaxis
immunization
toxoid
antitoxin

Study Tips

1. Good students learn to anticipate questions asked by their instructor in a course. As you continue to study microbiology, this is a skill that you can develop. Try writing some questions on the content of this chapter.

2. Encourage your instructor to invite a representative from the local health department to address your class. If this representative cannot visit your class, try to meet this individual and conduct a personal interview. It will offer you additional insights about the topic of preventing diseases.

3. Continue to answer the several kinds of questions at the end of each textbook chapter.

Preventing Disease/Epidemiology

A. Complete each of the following statements with the correct term.

   1. ________ is the name for a disease outbreak that affects many members of a population within a short time.

   2. A ___________ is an epidemic that spreads worldwide.

   3. A disease is _________ if it is always present in a population at about the same level.

   4. A __________ disease occurs only occasionally in a population.

B. Label each of the following statements as true or false. If false, correct it.

   1. Vital statistics are kept by almost all governments.

   2. Epidemiologists usually express health information as a rate.

   3. The prevalence rate is the rate of acquiring a disease during a certain period of time.

   4. The incidence rate is the rate of having a disease at any particular time.

5. If a disease currently infects 5 million people in the U.S., the incidence rate for that disease is one percent.

C. Match each of the following examples to the type of epidemiological study.

1. It tracks epidemic diseases.
2. It studies nosocomial infections.
3. It provides general information about diseases.
4. Koch conducted this type of study when seeking the cause of tuberculosis.
5. This investigates disease outbreaks.
6. This type of study revealed that the HIV retrovirus caused AIDS.
7. This study involves an infection introduced by a laparoscopy.
8. This study involves an infection resulting from a urinary catheterization.

A. descriptive
B. surveillance
C. field
D. hospital

Public Health

A. Short Answer

1. One method of disease prophylaxis is limiting people's exposure to pathogens. What is the other method?
2. How can a physician learn if a recent outbreak of rabies has occurred in a local community?
3. What is the national public health agency of the United States?
4. How is the disease typhoid fever spread in a human community?
5. Name the two general kinds of microorganisms that cause diarrhea.
6. What kinds of diseases were transmitted through milk consumption before the use of pasteurization?
7. In food preparation, how can trichinosis be prevented?

8. What is the single most important personal habit that can prevent the spread of disease?

9. Currently, how effective is DDT at controlling insect populations?

10. Are sexually transmitted diseases a problem only in developing countries?

11. What is the current, main preventive measure against the spread of AIDS?

12. How is the microorganism causing diphtheria transmitted?

Immunization

A. Short Answer

1. List the several characteristics that make a vaccine safe and effective.

2. List the several routes by which vaccines can be administered.

3. What is an acellular vaccine?

4. What is an attenuated vaccine?

5. What are the drawbacks to inactivated vaccines?

6. What were the first acellular vaccines?

7. A polysaccharide antigen can be conjugated with the protein diphtheria toxoid. How does this affect the antigen?

8. What is the current state of research of DNA vaccines?

9. What is passive immunization?

10. What is the drawback of passive immunization?

Discussion Questions

1. Why do you think some diseases are specific for humans and do not develop in other animal species?

2. Visit the Web site quoted in this chapter on DNA vaccines. What potential does this breakthrough hold for improving the prevention of diseases?

Multiple Choice: Review

1. Which kind of disease occurs only occasionally in a population?

   A. endemic
   B. epidemic
   C. pandemic
   D. sporadic

2. Pellagra develops in humans due to a

   A. bacterial infection.
   B. mineral deficiency.
   C. vitamin deficiency.
   D. viral infection.

3. A country has a human population of 150 million. 3 million people in this population are infected with a disease. The incidence rate for this disease is ________ percent.

   A. one
   B. two
   C. five
   D. six

4. Which type of epidemiological study provides general information about a disease?

   A. descriptive
   B. field
   C. hospital
   D. surveillance

5. Which type of epidemiological study investigates common source epidemics?

   A. descriptive
   B. field
   C. hospital
   D. surveillance

6. By urinary catheterization a device is placed in the

   A. bladder.
   B. kidney.
   C. ureter.
   D. urethra.

7. The term "prophylaxis" means

   A. immunization.
   B. infection.
   C. prevention.
   D. scrubbing.

8. Select the disease that is spread by an insect vector.

   A. cholera
   B. polio
   C. tetanus
   D. yellow fever

9. The development of Jenner's smallpox vaccine was an example of

   A. active immunization.
   B. passive immunization.

10. High titers of antibodies are administered to a person. This is an example of

   A. active immunization.
   B. passive immunization.

Answers

Preventing Disease/Epidemiology

A. 1. epidemic 2. pandemic 3. endemic 4. sporadic

B. 1. True
   2. True
   3. False; This defines the incidence rate.
   4. False; This defines the prevalence rate.
   5. False; The incidence rate is 5 million divided by 250 million or 2.0 percent

C. 1. B 2. D 3. A 4. A 5. C 6. C 7. D 8. D

Public Health

A. 1. immunization
   2. The physician can contact the local health department.
   3. United States Public Health Service
   4. It is spread by human sewage.

5. The microorganisms are the bacterium and virus
6. Examples included tuberculosis, typhoid fever, and scarlet fever.
7. It is prevented by cooking food adequately.
8. Hand washing is the most important preventive measure.
9. It is not currently effective.
10. They are a problem in both developing and developed countries.
11. Limited sexual exposure is the preventive measure
12. It is spread by respiratory droplets.

Immunization

A.
1. It must be free of side effects. It must be highly immunogenic. It must be administered appropriately.
2. The routes include orally, subcutaneously, and intramuscularly.
3. It contains only the antigen of the microorganism.
4. It is a genetically altered form of the microorganism.
5. The antigen stimulating immunity has been destroyed. Also, a microorganism is not available to continue multiplying to stimulate the immune system.
6. They were toxoids.
7. It makes the antigen more powerful.
8. It is an active field of research that will provide another means of immunization in the future.
9. By passive immunity the antibody is introduced into the subject. The subject's immune system is not stimulated to produce the antibody.
10. The effectiveness of passive immunity is short-lived.

Discussion Questions

1. The key to this answer resides in the different genetic blueprints in different species. This programs different immune systems with different capabilities.

2. Visit the Internet site. Try to compose a list of diseases that potentially will be treated through this technology.

Multiple Choice: Review

1. D 2. C 3. B 4. A 5. B 6. D 7. C 8. D 9. A 10. B

Chapter 21
Pharmacology

Principles of Pharmacology
- Drug Administration
- Drug Distribution
- Eliminating Drugs from the Body
- Side Effects and Allergies
- Drug Resistance
  - Natural Resistance
  - Acquired Resistance
    - Mechanisms
    - Genetics
    - Slowing Resistance
  - Drug Dosage
    - Disc-Diffusion Method
    - Broth-Dilution Method
    - Serum-Killing Power

Targets of Antimicrobial Drugs
- The Cell Wall
- Cell Membranes
- Protein Synthesis
- Nucleic Acids
- Folic Acid Synthesis

Antimicrobial Drugs
- Antibacterial Drugs
  - Penicillins
  - Cephalosporins
  - Sulfonamides
  - Aminoglycosides
  - Chloramphenicol
  - Tetracyclines
  - Erythromycin
  - Quinolones
  - Other Antibacterials
- Antimycobacterial Agents
  - Isoniazid
  - Rifampin
  - Ethambutol
- Antifungal Agents
  - Nystatin
  - Amphotericin B
  - Imidazole and Triazoles
  - Flucytosine
  - Giseofulvin

Antiparasitic Agents
Mebendazole
Metronidazole
Chloroquine
Antiviral Agents
Amantadine
Acyclovir
Ribavirin
Anti-HIV Agents
Reverse Transcriptase Inhibitors
Protease Inhibitors
Interferons
Summary

Key Terms

antibiotic
drug
chemotherapeutic agent
antimicrobial agent
external therapy
systemic therapy
intravenous administration
intramuscular administration
oral administration
side effect
narrow-sprectrum
broad-spectrum
penicillin-binding protein
combined therapy
disc-diffusion method
broth-dilution method
antimetabolite

Study Tips

1. Much of the information in this chapter deals with drugs and their physiological effects on microorganisms. Organize this information in the following table:

Drug	Effect

2. Try to relate the information in this chapter to other topics you have studied in biology. For example, how are the different types of drug administration related to the anatomy of the human body? How do different drugs inactivate different parts of the anatomy and metabolism of microorganisms?

3. Continue to answer the several kinds of questions at the end of the chapter. Ask your instructor to post an answer key for these questions after you have tried them.

Principles of Pharmacology

A. Label each of the following statements as true or false. If false, correct it.

1. A drug is any chemical that has a physiological effect on living things.
2. Antimicrobial agents are used to treat infectious diseases.
3. Most infections require external therapy.
4. The oral administration of a drug usually has rapid, efficient effects.
5. Most drugs enter cells by passing through the phospholipid part of the cell membrane.
6. The binding of drugs to proteins in the blood plasma usually stops their effective administration.
7. Drug metabolism occurs mainly in the liver.
8. Penicillin is excreted rapidly from the body through the kidneys.
9. Most antimicrobial drugs have side effects.
10. Penicillin is effective at treating viral infections.
11. Narrow-spectrum antibacterial drugs affect only a single microbial group.
12. Ampicillin is a narrow-spectrum antibiotic.
13. Broad-spectrum antibiotics can significantly alter the normal biota of the body.
14. Acquiring an enzyme that destroys a drug is the least common mechanism by which microbes become resistant to drugs.
15. Some resistant strains of *Neisseria gonorrhoeae* can pump tetracycline out of their cells.
16. An R factor encodes resistance to one specific antibiotic.

17. Generally in the environment, there is a rising tide of resistance to antibiotics by microorganisms.

18. By the Kirby-Bauer method, a clear halo develops around a disc impregnated with an antibiotic. The bacterium being tested is resistant to this antibiotic.

19. Disc-diffusion tests are relatively easy to perform.

20. The MIC shows the highest drug concentration that inhibits the growth of a bacterial population.

Targets of Antimicrobial Drugs

A. Match each drug to its principal target.

1. flucytosine	A. cell membranes of fungi
2. penicillin	B. cell wall
3. polymixin B	C. folic acid synthesis
4. sulfoanamides	D. nucleic acids
5. tetracycline	E. ribosome

Antimicrobial Drugs

A. Complete each of the following with the correct term.

1. The beta-lactamase-resistant penicillins are used to treat strains of the bacterium ________ ________.

2. The antibiotic ____________ of the penicillins is effective against treating *Pseudomonas aeruginosa*.

3. The cephalosporins are very effective against Gram - ____________ bacteria.

4. Among the families of antibiotics, the cephalosporins are very similar to the family of ____________.

5. The sulfonamides interfere with the ability of bacteria to synthesize ________ ________.

6. The aminoglycosides are effective against Gram - ________ bacteria.

7. Chloramphenicol can cause ________ anemia.

8. The tetracyclines interfere with the ability of the ribosome to accept the molecule ________ during protein synthesis.

9. Erythromycin is only effective against Gram - ________ bacteria.

10. The quinolones block DNA ________ in bacteria.

11. Vancomycin blocks ________ synthesis in bacteria.

12. In comparing cellular regions, *Mycobacterium tuberculosis* is an ________ pathogen.

13. Isoniazid is activated by the mycobacterial enzyme ________.

14. Rifampin is made by ________ *mediterranei*.

15. Nystatin is a member of the ________ family of antibiotics.

16. Amphotericin B disrupts the cell membranes of ________.

17. Griseofulvin is isolated from a fungus of the genus ________.

18. Metronidazole is effective against obligate ________ bacteria.

19. Amantadine is effective against the replication of ________.

20. Acyclovir is effective against DNA viruses of the ________ family.

21. Most reverse __________ inhibitors are nucleoside analogues.

22. ___________ are small glycoproteins that stimulate other cells to make antiviral agents.

Discussion Questions

1. How do the previously- studied units of cell structure and biochemistry contribute to the concepts of pharmacology?

2. Protease inhibitors are advertised frequently to treat human GI tract problems. What is their mode of action?

Multiple Choice: Review

1. A drug is introduced into the body by injection into the triceps brachii. This drug administration is

   A. EV.
   B. IV.
   C. IM.
   D. PO.

2. Most drugs are metabolized mainly by the

   A. liver.
   B. pancreas.
   C. small intestine.
   D. spleen.

3. Most penicillins are __________ drugs.

   A. broad-spectrum
   B. narrow-spectrum

4. *Streptococcus pneumoniae* is mainly a ________ pathogen.

   A. gastrointestinal
   B. nervous system
   C. respiratory
   D. urinary tract

5. By the disc-diffusion method, the more sensitive a bacterium is to an antibiotic, the ___________ the halo will be around the antibiotic disc.

   A. larger
   B. smaller

6. Streptomycin targets the __________ of bacterial cells.

   A. cell membrane
   B. cell wall
   C. nucleic acids
   D. ribosomes

7. Which antibiotics were the first to cure microbial infections after salvarsan?

   A. cephalosporins
   B. penicillins
   C. sulfoamides
   D. tetracyclines

8. Which antibiotic works against bacterial protein synthesis?

   A. aminoglycoside
   B. chloramphenicol
   C. penicillin
   D. tetracycline

9. Rifampin acts against

   A. Gram-positive bacteria only.
   B. Gram-negative bacteria only.
   C. Gram-negative and Gram-positive bacteria.
   D. viruses.

10. Select the antifungal agent that is a polyene.

   A. amphotericin B
   B. griesofulvin
   C. nystatin
   D. triazole

Answers

Principles of Pharmacology

A.
1. True
2. True
3. False; Most infections require systemic therapy.
4. False; Oral administration is slow and inefficient.
5. False; Most drugs are not lipid soluble.
6. False; Protein binding slows drug administration.
7. True
8. True
9. True
10. False; Viruses lack a cell wall and are resistant to penicillin.
11. True
12. False; Ampicillin affects Gram-negative and Gram-positive bacteria.
13. True

14. False; It is the most common mechanism.
15. True
16. False; It can encode resistance to several antibiotics.
17. True
18. False; This antibiotic stops microbial growth, as evidenced by the halo.
19. True
20. False; It shows the lowest drug concentration that inhibits bacterial growth.

Targets of Antimicrobial Drugs

A. 1. D 2. B 3. A 4. C 5. E

Antimicrobial Drugs

A.
1. *Staphylococcus aureus*
2. carbenicillin
3. positive
4. penicillins
5. folic acid
6. negative
7. aplastic
8. aminoacyl-tRNA
9. positive
10. replication
11. peptidoglycan
12. intracellular
13. peroxidase
14. *Streptomyces*
15. polyene
16. fungi
17. *Penicillium*
18. anaerobic
19. viruses
20. herpes
21. transcriptase
22. interferons

Discussion Questions

1. Most of the antimicrobial drugs target the cell wall, cell membrane, genetic structure, and ribosomes of microbial cells. In order to understand drug action, you must understand cell structures and functions. Each kind of drug has a unique chemical structure that accounts for its mode of action.

Chapter 21

2. Review the structure of enzymes. Most of the makeup of an enzyme is protein.

Multiple Choice: Review

1. C 2. A 3. B 4. C 5. A 6. D 7. C 8. D 9. C 10. C

Chapter 22
Infections of the Respiratory System

- The Respiratory System
  - Structure and Function
  - Defenses and Normal Biota: A Brief Review
  - Clinical Syndromes
- Upper Respiratory Infections
  - Bacterial Infections
    - Epiglottitis
    - Streptococcal Pharyngitis
      - The Clinical Syndrome
      - Complications and Changing Virulence
    - Diphtheria
      - The Clinical Syndrome
      - Epidemiology and Prevention
    - Viral Infections
      - The Common Cold
        - Rhinoviruses
        - Other Causes of the Common Cold
- Lower Respiratory Infections
  - Bacterial Infections
    - Pneumococcal Pneumonia
      - The Organism
      - The Clinical Syndrome
      - Treatment and Prevention
      - Other Acute Bacterial Pneumonias
      - Mycoplasmal Pneumonia
      - Chlamydial Pneumonia
      - Q Fever
      - Legionellosis
      - Pertussis (Whooping Cough)
      - Tuberculosis
        - The Clinical Syndrome
        - Prevention and Treatment
        - TB in the 1990's
      - Bovine Tuberculosis
    - Viral Infections
      - Influenza (Flu)
        - Epidemiology
        - Clinical Syndromes
        - Prevention and Treatment
      - Croup
      - Bronciolitis
      - Hantavirus Pulmonary Syndrome

Fungal Infections
Histoplasmosis
Coccidiodomycosis
Blastomycosis
Pneumocyctis Pneumonia
Summary

Key Terms

nasal cavity
adenoids
pharynx
trachea
larynx
pleura
epiglottis
pneumonia
scarlet fever
rheumatic fever
rhinorrhea
quelling reaction
sepsis
filamentous hemagglutinin
pertussis toxin
tuberculin skin testing

Study Tips

1. Most of the information in this chapter relates respiratory diseases to causative microorganisms. Organize this information by composing the following table:

Disease	Microorganism

2. In addition to the information presented at the beginning of the chapter, study the human anatomy of the respiratory system in more depth. It will help you to understand the microorganisms and diseases discussed for the balance of the chapter.

3. An important emphasis of your study is to relate microbial activity to the different lines of defense that protect the respiratory system. This approach will help you to understand the various mechanisms of disease. For example, a microorganism can infect the body by overcoming the mucociliary system of the respiratory tract.

4. Continue to answer the several kinds of questions at the end of the chapter.

The Respiratory System

A. Complete each of the following statements with the correct term describing the respiratory tract.

1. After the trachea, exhaled air enters the _________ next.
2. The adenoids are located in the ____________.
3. From the primary bronchus, exhaled air enters the _________ next.
4. Oxygen and carbon dioxide diffuse across the walls of the __________.
5. The ________ and __________ lack the mucociliary system.

B. Short Answer

1. Describe the conditions of the upper respiratory tract that make it favorable for colonization of microbes.
2. What is rhinitis?
3. What is otitis media?
4. What is tachypnea?
5. How are the alveoli affected by pneumonia?

Upper Respiratory Infections

A. The following statements are about bacterial infections. Label each of the following statements as true or false. If false, correct it.

1. The epiglottis is prone to infection by *Haemophilus influenzae*.
2. Strains of *Haemophilus influenzae* can invade only human tissues.
3. Ampicillin is effective at treating respiratory infections caused by *Haemophilus influenzae*.
4. Alpha-hemolytic strains of *Streptococcus pyrogenes* can lyse red blood cells.
5. Humans are the only natural reservoir for group A streptococci.
6. Streptococcal pharyngitis usually lasts several weeks in an infected person.
7. Type B is the most virulent toxin causing streptococcal pharyngitis.

8. Rheumatic fever causes inflammation of many organs of the body.

9. Glomerulonephritis is an upper respiratory tract infection.

10. *Corynebacterium diphtheriae* is Gram-negative and rod-shaped.

11. The diphtheria toxin is an endotoxin.

12. The diphtheria antitoxin will neutralize the diphtheria toxin already formed in the body.

13. In the development of membranous pharyngitis, the lymph nodes of the neck swell.

14. Toxin-producing strains of *C. diphtheriae* are now rare.

B. Viral Infections

The following statements are about viral infections. Label each statement as true or false. If false, correct it.

1. The common cold is caused by a virus.

2. Rhinoviruses are DNA viruses.

3. The only reservoir for rhinoviruses is human beings.

4. By IgA antibody production, people infected by a particular rhinovirus are protected against it for two months or less after the infection.

5. Antibiotics are useless at treating the common cold.

6. Interferon treatment prevents about 70 percent of common colds.

Lower Respiratory Infections

A. The following statements are about bacterial infections. Complete each statement with the correct term or terms.

1. Pneumoncoccal pneumonia is caused by the bacterium ________ ________.

2. The swelling of the capsule in a pneumonia-causing bacterium is called a __________ reaction.

3. In the respiratory system, aspirating is the insertion of a needle into the ________ or ________.

4. A consolidated region of the lung is filled with ________.

5. By pneumococcal ________, pneumococci multiply in the blood.

6. __________ is the vaccine used for immunization against pneumoncoccal pneumonia.

7. ________ __________ is the causative agent of mycoplasmal pneumonia.

8. Mycoplasmal pneumonia is usually treated with the antibiotic ________.

9. Chlamydiae are obligate ________ ________.

10. *C. psittaci* commonly affects all kinds of ________.

11. Q fever is caused by the bacterium ________ ________.

12. Q fever is usually treated with the antibiotic ________.

13. *L. pneumophila* multiplies intracellularly in the host as macrophages cannot _________ it.

14. The antibiotic ____________ is used to treat legionellosis.

15. Another name for pertussis is _________.

16. *B. pertussis* produces a protein called filamentous ________, which attaches to susceptible cells.

17. _________ is the leading killer among infectious diseases.

18. *M. tuberculosis* grows best where oxygen concentrations are _________.

19. The doubling time for *M. tuberculosis* is about ________ hours.

20. Calcified tubercles caused by *M. tuberculosis* are called ________ complexes.

21. Tuberculosis ___________ infects the central nervous system.

22. BCG vaccination interferes with tuberculin ________ ________.

23. During the 1980s the rate of tuberculosis cases has been ________.

24. *M. bovis* infects ________.

25. *M. avium* causes a form of the disease ________.

B. The following statements are about viral infections. Label each statement as true or false. If false, correct it.

1. Infection by the influenza virus stimulates the production of antibodies that prevent reinfection.

2. Influenza viruses are named by the particular antigens they synthesize.

3. The influenza virus can be transmitted directly from birds to humans.

4. Tracheobronchitis weakens the mucociliary system.

5. Vaccination protects about 20 percent of recipients from influenza.

6. Amantadine is effect as an immunization in preventing influenza A.

7. Croup is most common in adults.

8. Croup epidemics occur during the fall.

9. RSV infection occurs during the late winter and early spring.

10. The virus for the hantavirus pulmonary syndrome occurs mainly in rats.

C. The following statements are about fungal infections. Label each statement as true or false. If false, correct it.

1. Fungal infections commonly infect the lower respiratory tract.

2. Histoplasmosis occurs mainly in the United States.

3. About 50 percent of the people infected with the histoplasmosis fungus become ill.

4. Most people infected with coccidiodomycosis suffer only a mild respiratory illness.

5. Blastomycosis symptoms can resemble pulmonary tuberculosis.

6. *Pneumocystis carinii* is a protozoan.

Discussion Questions

1. How does the information of this chapter draw on concepts you have learned earlier in the text?

2. Survey the upcoming chapters in the text. How will the concepts in this chapter help you to comprehend the important ideas in these upcoming chapters.

Multiple Choice: Review

1. Select the incorrect association.

    A. larynx/voicebox
    B. pleura/lines bronchi
    C. tonsil/lymphoid tissue
    D. trachea/windpipe

2. By bronchitis, the bronchi

    A. shrink and produce a thin mucus.
    B. shrink and produce a thick mucus.
    C. swell and produce a thin mucus.
    D. swell and produce a thick mucus.

3. The lettered groups of *Streptococcus pyrogenes* refer to the type of

    A. Gram-staining reactions of the bacteria.
    B. location of infection in the human body.
    C. polysaccharide antigen present on cell surfaces.
    D. tendency to produce hemolysis in infected cells.

4. Rheumatic fever causes the most life-threatening damage to the

    A. heart valves.
    B. kidneys.
    C. joints.
    D. skeletal muscles.

5. The diphtheria toxin is a

    A. two-subunit endotoxin.
    B. two-subunit exotoxin.
    C. three-subunit endotoxin.
    D. three-subunit exotoxin.

6. The common cold is caused by a

    A. bacterium.
    B. fungus.
    C. protozoan.
    D. virus.

7. The coronavirus is a

   A. large-sized DNA virus.
   B. large-sized RNA virus.
   C. medium-sized DNA virus.
   D. medium-sized RNA virus.

8. About ________ percent of healthy adults harbor *S. pneumoniae*.

   A. five
   B. ten
   C. twenty
   D. fifty

9. Each is a genus causing pneumonia except

   A. *Haemophilus*.
   B. *Klebsiella*.
   C. *Pseudomonas*.
   D. *Staphylococcus*.

10. Q fever is caused by a

   A. fungus.
   B. protozoan.
   C. rickettsia.
   D. virus.

Answers

The Respiratory System

A. 1. larynx 2. nasopharynx 3. trachea 4. alveoli 5. bronchioles and alveoli

B. 1. It is warm, moist, and nutrient-rich.
   2. It is nasal inflammation.
   3. It is the infection of the middle ear.
   4. It is rapid breathing.
   5. They become filled with fluid.

Upper Respiratory Infections

A. 1. True
   2. True

3. False; This antibiotic is useless for this treatment.
4. False; Beta-hemolytic strains lyse red blood cells.
5. True
6. False; Most people infected with this recover in a few days.
7. False; Type A is the most virulent toxin.
8. True
9. False; It is an inflammation of the kidney.
10. False; It is Gram-positive.
11. False; It is an exotoxin.
12. False; It will only stop further production and transmission of the disease.
13. True
14. True

B. 1. True
2. False; They are single-stranded RNA viruses.
3. True
4. False; They are protected for 18 months.
5. True
6. False; This treatment prevents about 40 percent of common colds.

Lower Respiratory Infections

A. 1. *Streptococcus pneumoniae*
2. quelling
3. lungs or trachea
4. fluid
5. sepsis
6. Pneumovax
7. *Mycoplasma pneumoniae*
8. tetracycline
9. intracellular parasites
10. birds
11. *Coxiella burnetii*
12. tetracycline
13. phagocytize
14. erythromycin
15. whooping cough
16. hemagglutinin
17. tuberculosis
18. high
19. twenty
20. Ghon
21. meningitis
22. skin testing
23. increasing

24. cows
25. tuberculosis

B. 1. True
2. False; They are named by abbreviations, numbers, or the place first identified.
3. False; They can exchange it with pigs that infect humans.
4. True
5. False; It protects about 70 percent of the recipients.
6. True
7. False; It is most common in toddlers.
8. True
9. True
10. False; It occurs mainly in long-tailed deer mice.

C. 1. True
2. False; It occurs worldwide.
3. False; Only about one percent of the people infected become ill.
4. True
5. True
6. False; It is a fungus.

Discussion Questions

1. This chapter incorporates earlier concepts of cell morphology, biochemistry, metabolism, genetics, and taxonomy.

2. This chapter offers ideas for mechanisms of disease and immune responses that you will study in the infections of other systems of the body.

Multiple Choice: Review

1. B 2. D 3. C 4. A 5. B 6. D 7. D 8. B 9. C 10. C

Chapter 23
Infections of the Digestive System

- The Digestive System
  - Structure and Function
  - Defenses and Normal Biota: A Brief Review
  - Clinical Syndromes
- Infections of the Oral Cavity and Salivary Glands
  - Bacterial Infections
    - Dental Caries
    - Periodontal Disease
  - Viral Infections
    - Mumps
- Infections of the Intestinal Tract
  - Bacterial Infections
    - Shigellosis
      - Transmission and Epidemiology
      - Clinical Syndrome and Treatment
    - Typhoid Fever
    - Salmonellosis
    - *Escherichia coli* Diarrhea and Hemolytic-Uremic Syndrome
    - Cholera
      - Symptoms and Treatment
      - Recent Epidemiology
    - *Vibrio parahaemolyticus* Gastroenteritis
    - *Yersinia enterocolitica* Enterocolitis
    - Campylobacteriosis
    - *Helicobacter pylori* Gastritis, Peptic Ulcer Disease
    - *Clostridium difficile* Diarrhea
    - Staphylococcal Food Poisoning
      - Other Food Poisonings
        - *Bacillus cereus*
        - *Clostridium perfringens*
        - *Clostridium botulinum*
    - Viral Infections
      - Rotavirus Gastroenteritis
      - Norwalk Agents Gastroenteritis
    - Protozoal Infections
      - Amoebic Dysentery
      - Giardiasis
      - Balantidiasis
      - Cryptosporidiosis
    - Helminthic Infections
      - Pinworm
      - *Ascaris lumbricoides*

Hookworm
Strongyloides stercoralis
Whipworm
Trichinella spiralis
Tapeworms
Infections of the Liver
Viral Infections
Hepatitis A Virus (HAV)
Hepatitis B Virus (HBV)
Hepatitis C Virus (HCV)
Hepatitis Delta Agent
Hepatitis E Virus (HEV)
Helminthic Infections
Liver Fluke Infections
Summary

Key Terms

peristalsis
esophagus
chyme
duodenum
jejunum
ileum
large intestine
gallbladder
gastritis
gastroenteritis
hepatitis
cariogenic
periodontum
gingivitis
exanthems
orchitis
shiga toxin
dehydration
infectious hepatitis
hepatocytes

Study Tips

1. Most of the information in this chapter relates digestive diseases to causative microorganisms. Organize this information by composing the following table.

Disease	Microorganism

2. Study the anatomy of the digestive system in more depth by consulting texts on human anatomy and physiology. Learning more about these topics will help you to understand the microorganisms and diseases discussed throughout the chapter.

3. Does your biology lab have a human torso model? Study it as another aid to enhance your understanding of the human anatomy relevant to this chapter and other chapters.

4. Study prepared slides of some of the causative microorganisms described in this chapter. Pay particular attention to the cell morphology and Gram-staining reactions described.

The Digestive System

A. Complete each of the following statements with the correct term or terms.

1. ________ is the intestinal movement that propels food through the digestive tract.
2. The secretion of bile is specific for the breakdown of lumps of ________ in the diet.
3. The ________, ________, and ____________ are regions of the small intestine.
4. The pancreas does not have a normal biota due to __________ and __________ defenses.
5. ______________ is the inflammation of the stomach.
6. Colitis mainly affects the ________ of the digestive tract.
7. The term "periodontal" means ________.
8. Hepatitis involves damage to the ________.

Infections of the Oral Cavity and Salivary Glands

A. Complete each of the following statements with the correct term or terms.

1. *S. mutans* attaches to the ________ of the oral cavity.

2. *S. mutans* is ________, meaning that it is caries-producing.

3. The fermentation of sucrose in the oral cavity produces ________ acid.

4. Incorporation of the mineral __________ into tooth enamel makes the teeth more resistant to caries.

5. The root of a tooth is normally covered with ________.

6. ________ is the inflammation of the gums.

7. The primary causative bacterium of gingivitis is ________.

8. __________ are skin rashes.

9. Mumps is caused by a ________ microorganism.

10. Since the MMR vaccine the incidence of mumps in the U.S. have decreased ________ percent.

Infections of the Intestinal Tract

A. Label each of the following statements as true or false. If false, correct it.

1. *Shigella* species are Gram-negative.

2. *Shigella* species cause microabscesses in the colon.

3. Shigellosis has been nearly eradicated in the United States.

4. More than a 15 percent loss of human body weight by dehydration is usually fatal.

5. Typhoid fever is caused by a species of *Streptococcus*.

6. Typhoid fever has become a rare disease in developed countries.

7. Salmonellosis is a true infection caused by bacteria multiplying in the intestine.

8. Salmonellosis is usually treated with antibiotics.

9. *E. coli* is an invasive pathogen to the human intestine.

10. The heat labile toxin causing *E. coli* diarrhea is similar to the toxin that causes cholera.

11. Enteroinvasive strains of *E. coli* cause a dysentery almost identical to shigellosis.

12. Cholera is caused by a species of *Vibrio*.

13. Chlorea is caused by the retention of chloride ions in the intestinal tract.

14. The fluids of ORT contain glucose and sodium.

15. *V. parahaemolyticus* is a Gram-positive rod.

16. A species of *Yersinia* causes enterocolitis.

17. *Campylobacter* species are slightly curved, Gram - negative rods.

18. Stomach ulcers are caused by a virus.

19. Iatrogenic diarrhea is caused by a species of *Staphylococcus*.

20. Strains of *S. aureus* that cause food poisoning produce a heat-stable protein enterotoxin.

21. *B. cereus* lives in soil and water.

22. *C. perfringens* causes botulism.

23. *C. botulinum* causes gas gangrene.

24. A rotavirus can cause gastroenteritis.

25. About one-third of all gastroenteritis outbreaks are caused by Norwalk agents.

B. Match each protozoan disease to its correct description.

1.	amebic dysentery	A.	It causes severe problems for AIDS patients.
2.	giardiasis	B.	It is caused by a ciliated protozoan.
3.	balantidiasis	C.	It occurs as a trophozoite.
4.	cryptosporidiosis	D.	It is caused by a protozoan with genus name *histolytica*.

C. Helminth Infections/Short Answer

1. Which helminth causes the most common helminthic infection?

2. Where does *Ascaris lumbricoides* mature in the lungs?

3. How do hookworms penetrate the human body?

4. Where do the eggs of *Strongyloides stercoralis* hatch in the human body?

5. What is the genus name for the whipworm?

6. How are humans infected in cysticercosis?

Infections of the Liver

A. Complete each of the following statements with the correct term.

1. The hepatitis A virus contains single-stranded ________.

2. _________ are liver cells.

3. Hepatitis A is distinguished by the presence of ________ antibodies.

4. The ________ virus is the best-known virus to cause human cancer.

5. Worldwide there are more than _________ million carriers of HBV.

6. Heptatitis B will be prevented by the administration of an effective ________.

7. HCV is transmitted by sexual contact or ________.

8. Chronic hepatitis can progress to the disease ________ of the liver.

9. The nucleic acid in HEV is ________.

10. *Fasciola hepatica* is the _________ liver fluke.

Discussion Questions

1. What concepts learned in earlier chapters tie into the concepts of infections of the human digestive tract?

2. What information from other branches of science relate to the key concepts of this chapter?

Multiple Choice: Review

1. Bile is made by the __1__ and stored mainly in the __2__.

   A. 1 - gallbladder, 2 - liver
   B. 1 - liver, 2 - gallbladder

2. Dental caries is caused by a species of the genus

   A. *Bacillus.*
   B. *Escherichia.*
   C. *Staphylococcus.*
   D. *Streptococcus.*

3. Gingivitis is caused by a species of the genus

   A. *Bacillus.*
   B. *Bacteroides.*
   C. *Staphylococcus.*
   D. *Streptococcus.*

4. Shigellosis is caused by a

   A. Gram-negative coccus.
   B. Gram-negative rod.
   C. Gram-positive coccus.
   D. Gram-positive rod.

5. During dehydration, in the intestinal tract

   A. chloride ions follow the loss of water.
   B. water follows the loss of chloride ions.

6. Select the correct statement about salmonellosis.

   A. It is a true infection.
   B. It is no longer a major health concern.
   C. It is treated successfully with antibiotics.
   D. It is usually not self-limiting.

7. Diarrhea caused by *E. coli* is treated best with the antibiotic

   A. cephalosporin.
   B. ciprofloxacin.
   C. erythromycin.
   D. tetracycline.

8. Cholera is caused by a species of the genus

   A. *Clostridium.*
   B. *Pseudomonas.*
   C. *Staphylococcus.*
   D. *Vibrio.*

9. Giardiasis is caused by a

   A. bacterium.
   B. fungus.
   C. protozoan.
   D. virus.

10. Select the cancer-causing virus.

   A. HAV
   B. HBV
   C. HCV
   D. HEV

Answers

The Digestive System

A. 1. peristalsis
   2. fat
   3. duodenum, jejunum, ileum
   4. nonspecific, specific
   5. gastritis
   6. colon
   7. around the teeth
   8. liver

Infections of the Oral Cavity and Salivary Glands

A. 1. teeth
   2. cariogenic
   3. lactic
   4. fluoride
   5. cementum
   6. gingivitis
   7. *B. gingivalis*
   8. enanthems
   9. virus
   10. 97

Infections of the Intestinal Tract

A. 1. False; They are Gram-positive.

2. True
3. False; Thousands of cases are reported annually.
4. True
5. False; It is caused by *Salmonella typhi*.
6. True
7. True
8. False; This is contraindicated.
9. False; It is a normal biota of the human intestine.
10. True
11. True
12. True
13. False; It is stimulated by the loss of chloride ions from the intestinal tract.
14. True
15. False; It is Gram-negative.
16. True
17. True
18. False; They are caused by a bacterium.
19. False; It is medically induced diarrhea.
20. True
21. True
22. False; It causes gas gangrene.
23. False; It causes botulism.
24. True
25. True

B. 1. D 2. C 3. B 4. A

C.
1. pinworm
2. alveoli
3. They penetrate the human skin.
4. human intestine
5. *Trichuris*
6. Larvae encyst in human tissues.

Infections of the Liver

A. Complete each of the following statements with the correct term.

1. RNA
2. hepatocytes
3. antiviral
4. HBV
5. 300
6. vaccine
7. contaminated blood

8. cirrhosis
9. RNA
10. sheep

Discussion Questions

1. Review bacterial cell morphology and the Gram stain. Concentrate on adaptations that make microbes successful at overcoming the lines of defense in the human body.

2. Study more about the structure and function of the human body. What are the layers of the tooth? What are the four layers of the GI tract? Where does absorption occur? How does peristalsis occur?

Multiple Choice: Review

1. B 2. D 3. B 4. B 5. B 6. A 7. B 8. D 9. C 10. B

Chapter 24
Infections of the Genitourinary System

- The Genitourinary System
- Structure of the Urinary System
  - Defenses and Normal Biota: A Brief Review
  - Clinical Syndromes
- Urinary Tract Infections
  - Diagnosis
  - Lower versus Upper Tract Infections
  - Treatment
  - Bacteria that Cause UTIs
  - Leptospirosis
- Sexually Transmissible Diseases (STDs)
  - Bacterial Infections
    - Gonorrhea
      - Clinical Syndromes
      - Prevention and Treatment
    - Syphilis
      - Diagnosis
      - Prevention and Treatment
    - Chlamydia
    - Lymphogranuloma Venereum
    - Nongonoccal Urethritis
    - Chancroid
    - Granuloma Inguinale
  - Viral Infections
    - Herpes
      - Clinical Syndrome
      - Prevention and Treatment
    - Genital Warts
    - Other Viral STDs
- Infections of the Female Reproductive Tract
  - Bacterial Infections
    - *Gardnerella vaginitis*
    - Toxic Shock Syndrome
    - Pelvic Inflammatory Disease
    - Infections after Childbirth or Abortion
  - Fungal and Protozoal Infections
    - Candidiasis
    - Trichmoniasis
- Infections of the Male Reproductive Tract
- Infections Transmitted from Mother to Infant
  - Bacterial Infections
    - Listeriosis

Key Terms

bladder
urethra
gamete
testes
prostate gland
semen
ovary
uterus
cervix
vagina
embryo
sexually transmissible disease
cystitis
urethritis
pyelonephritis
urinary tract infection
vaginitis
salpingitis
oophoritis
pelvic inflammatory disease
perinatal
kidney stone
gonorrheal endotoxin
purulent
nontreponemal test
syphilis
snuffles
lymphogranuloma venereum
granuloma inguinale
primary infection
puerperal sepsis

Study Tips

1. Much of the information in this chapter relates genitourinary diseases to causative microorganisms. Organize this information by composing the following table:

Disease	Microorganism

2. Consult additional sources to study the anatomy and physiology of the urinary and reproductive systems in more depth. This will promote your understanding of the infections of the urinary and reproductive tracts.

3. Your lab experiences can also facilitate understanding of the concepts in this chapter. Are there models available in your lab showing the urinary and reproductive systems? Have you studied some of the microorganisms described in this chapter? Start by reviewing their cell morphology and Gram-stain reactions.

4. Continue to answer the questions at the end of the chapter. After answering these questions independently, ask your instructor to post an answer key for these questions.

The Genitourinary System/Structure and Function of the Urinary System

A. Complete each of the following statements with the correct term or terms.

1. The _________ and _________ tracts merge in the male.

2. The glomerulus is a cluster of ________ _________ in the kidney.

3. Urine leaves the kidney through the _________ of the urinary tract.

4. Urine is stored in the ________.

5. In the male, the _________ of the urinary tract is a longer structure than in the female.

Structure and Function of the Reproductive System

A. Complete each of the following statements with the correct term or terms.

1. A _________ is a ovum or a sperm.

2. In the male the ________ is carried in the scrotum.

3. ___________ is a mixture of sperm cells and seminal fluid.

4. The __________ is a necklike extension of the uterus.

5. The internal lining of the urinary tract consists of tightly-joined ________ cells.

6. _________ is the infection of the bladder.

7. _________ is an infection of the lining of uterus.

8. __________ is an infection of the fallopian tubes.

Urinary Tract Infections

A. Label each of the following statements as true or false. If false, correct it.

1. All major pathogens of the urinary tract are bacteria.

2. Urinary tract infections are extremely common.

3. Males are more vulnerable than females to UTIs.

4. A normal urine sample contains very few bacteria.

5. The presence of leukocyte esterase in the urine indicates the breakdown of white blood cells in the urine.

6. *Staphylococcus aureus* is the most common pathogen of the urinary tract.

7. *Leptospira interrogans* usually enters the human body by a break in the serous membranes of the skin.

8. *L. interrogans* is a large bacterium and easy to diagnose.

Sexually Transmissible Diseases (STDs)

A. Label each of the following statements as true or false. If false, correct it.

1. Gonorrhea is caused by a species of *Neisseria.*

2. The rate of outbreaks of gonorrhea has continued to increase over the last 20 years.

3. The presence of Gram-positive diplococci in an urethral discharge indicates the presence of *N. gonorrhoeae* in the urinary tract.

4. The presence of capsules on gonococcal cells promotes adherence of these pathogens to the epithelial of the urinary tract.

5. *N. gonorrhoeae* infects only humans.

6. Gonococcal infections can lead to the development of arthritis.

7. Gonococci can become penicillin-resistant through chromosomal mutations or by acquiring a plasmid.

8. Recently, in the United States the rate of syphilis outbreaks has increased markedly.

9. *T. pallidum* infects only human beings.

10. Secondary syphilis begins 12 to 14 weeks after the chancre appears.

11. Latent syphilis develops after late syphilis.

12. Gummas develop in the cardiovascular system.

13. Chlamydia is the leading cause of female infertility.

14. Chlamydia can be cured with erythromycin.

15. Lymphogranuloma venereum develops in the cervical lymph nodes.

16. *C. granulomatis* is a Gram-negative rod.

17. Herpes simplex is the most commonly sexually transmitted pathogen in industrialized countries.

18. HSV viral genes are expressed in the host cells during the latent infection.

19. Neonatal herpes spreads to the brain and other internal organs.

20. The administration of acyclovir can cure genital herpes.

21. Genital warts is commonly reported.

22. HPV can lead to cancer.

Infections of the Female Reproductive Tract

A. Complete each of the following with the correct term.

1. *G. vaginalis* is a Gram __________ bacterium.

2. *G. vaginitis* can be diagnosed if ____________ cells are sloughed off from the vagina.

3. Toxic shock syndrome is caused by strains of _______ ________.

4. PID increases the risk for a pregnancy that is ________.

5. Endometritis is particularly likely to occur after ___________ or ________.

6. ___________ streptococci are one causative group for endometritis.

7. *Candida albicans* is a yeast-like ________.

8. *T. vaginalis* is a ________ protozoan.

Infections of the Male Reproductive Tract

A. Complete each of the following with the correct term.

1. __________ is the inflammation of the male reproductive tract.

2. _________ _________ is the most common causative agent of nongonococcal urethritis in males.

Infections Transmitted from Mother to Infant

A. Label whether each of the following is caused by a bacterium or a virus.

1. cytomegalic inclusion disease

2. listeriosis

B. Complete each of the following with the correct term or terms.

1. *L. monocytogenes* is an opportunistic ________.

2. Group B streptococci is found in the ________, _________, and vagina.

3. CMV is transmitted by the exchange of _________ and ________.

4. __________ is an effective drug for treating CMV.

Discussion Questions

1. The human urinary and reproductive tracts offer the perfect environment for the growth of numerous microorganisms. What are some of the optimal conditions it offers?

2. Your knowledge of microbiology is growing. How does the information you have learned promote the current concepts you are studying?

Multiple Choice: Review

1. Infection of the fallopian tubes is

   A. endometritis.
   B. oophoritis.
   C. salpingitis.
   D. vaginitis.

2. The pathogen most commonly infecting the female reproductive tract is

   A. *B. subtilis*.
   B. *E. coli*.
   C. *P. aeruginosa*.
   D. *S. aureus*.

3. 90 percent of *E. coli* strains are sensitive to

   A. cefalexin.
   B. erythromycin.
   C. streptomycin.
   D. tetracyline.

4. Select the incorrect statement about *L. interrogans*.

   A. It is a member of the spirochete family.
   B. It is a small bacterium.
   C. It is difficult to diagnose.
   D. It is reported frequently in the U.S.

5. Select the incorrect statement about *N. gonorrhoeae*.

   A. It is a fragile pathogen.
   B. It is a highly adapted pathogen.
   C. It is easily killed when exposed to sunlight.
   D. It is Gram-positive.

6. Skin lesions appear during _________ syphilis.

   A. primary
   B. secondary
   C. tertiary
   D. quaternary

7. What is the leading cause of female infertility in the U.S.?

   A. chlamydia
   B. gonorrhea
   C. herpes
   D. syphilis

8. HSV is a(n)

   A. DNA virus that affects the heart.
   B. DNA virus that affects neurons.
   C. RNA virus that affects the heart.
   D. RNA virus that affects neurons.

9. Genital warts is caused by a

   A. bacterium.
   B. fungus.
   C. protozoan.
   D. virus.

10. TSST is caused by a species of

    A. *Escherichia.*
    B. *Pseudomonas.*
    C. *Staphylococcus.*
    D. *Streptococcus.*

Answers

The Genotiurinary System/Structure and Function of the Reproductive System

A. 1. urinary, reproductive
   2. blood vessels
   3. ureter
   4. bladder
   5. urethra

Structure and Function of the Reproductive System

A. 1. gamete
2. testis
3. semen
4. cervix
5. epithelial
6. cystitis
7. endometritis
8. salpingitis

Urinary Tract Infections

A. 1. True
2. True
3. False; Females are more vulnerable.
4. True
5. False; It indicates that white blood cells are present in the urine.
6. False; *E. coli* is the most common pathogen of the urinary tract.
7. False; It enters by a break in the mucous membranes.
8. False; It is small and difficult to detect.

Sexually Transmissible Diseases (STD's)

A. 1. True
2. False; It has declined steadily since then.
3. False; Finding Gram-negative diplococci indicates its presence.
4. False; Pili promote the adherence of this microbe.
5. True
6. True
7. True
8. False; It has declined steadily over the last ten years.
9. True
10. False; It occurs 6 to 8 weeks after a chancre appears.
11. False; It develops before late syphilis.
12. False; They develop in bones or skin.
13. True
14. True
15. False; It develops in the lymph nodes of the groin.
16. False; It is a Gram-positive rod.
17. True
18. False; The viral genes are not expressed in this stage.
19. True
20. False; It cannot be cured.

21. False; It is not a reportable disease.
22. True

Infections of the Female Reproductive Tract

A. 1. positive
   2. clue
   3. *Staphylococcus aureus*
   4. ectopic
   5. childbirth, abortion
   6. anaerobic
   7. fungus
   8. flagellated

Infections of the Male Reproductive Tract

A. 1. urethritis
   2. *Chlamydia trachomatis*

Infections Transmitted from Mother to Infant

A. 1. virus
   2. bacterium

B. 1. pathogen
   2. throat, gastrointestinal tract
   3. saliva and infected blood
   4. gangiclovir

Discussion Questions

1. These tracts provide a warm, moist environment in the absence of light. The physical conditions of temperature and moisture are optimal for many microbes.

2. Review the details of cell morphology. Also review the patterns of anaerobic and aerobic metabolism.

Multiple Choice: Review

1. C 2. B 3. A 4. D 5. D 6. B 7. A 8. B 9. D 10. C

Chapter 25
Infections of the Nervous System

Chapter 25

Key Terms

central nervous system
peripheral nervous system
nerve
meninges
dura mater
arachnoid
pia mater
cerebrospinal fluid
lumbar puncture
blood-brain barrier
meningitis
encephalitis
myelitis
neurotoxin
intoxication
protease
invasive
meningococcal prophylaxis
aseptic meningitis
chronic meningitis
tetanus
tetanospasmin
trismus
tetanus antitoxin
botulism
botulinum toxin
rabies
prodromal phase
excitation phase
hydrophobia
paralytic phase
arbovirus
subclinical
African sleeping sickness

Study Tips

1. Answer the several kinds of questions at the end of the chapter. After answering them, discuss your answers with your instructor and classmates.

2. New vocabulary is a major part of this chapter. Organize the key terms as follows:

   Term Definition

After studying the text, define each term in your own words.

3. Can you tie in information from other courses to the content of this chapter? Start by outlining the human central and peripheral nervous system. Describe the functions of the neuron and neuroglial cells. What is the composition of the blood? What is the structure and function of the respiratory system?

4. These concluding chapters in the text draw on basic information that you have already studied. Can you find this information in the text? Begin with basic chemistry (Chapter 2), prokaryotic cells (Chapter 4), and viruses (Chapter 13). A review of these chapters will help you to understand the content of Chapter 25.

The Nervous System

A. Label each of the following statements as true or false. If false, correct it.

1. Neuroglia cells send impulses.
2. The spinal cord is part of the CNS.
3. The dura mater is the innermost meningeal layer.
4. The CSF circulates between the dura mater and arachnoid mater.
5. Glucose can pass through the blood brain barrier.
6. Some antibiotics cannot pass through the blood brain barrier.
7. Myelitis is the inflammation of the meninges.
8. Encephalitis is inflammation of brain tissue.

Infections of the Meninges: Bacterial Causes

A. Select the correct statements about *Neisseria meningitidis*.

1. It is a Gram-positive pathogen.
2. Its only natural reservoir is the human body.
3. It is usually grown on a very rich nutrient medium.
4. It is transmitted from person to person by respiratory droplets.

5. It adds iron to transferrin in the human body.

6. One of the symptoms of its disease is producing petechiae on the body.

7. It cannot destroy IgA antibodies.

8. It usually affects brain function.

B. Select the correct statements about *Haemophilus influenzae*.

1. It has a polysaccharide capsule.

2. It infects the urinary tract.

3. Tetracycline is commonly prescribed to treat carriers of this microbe.

4. Until 10 years ago it caused 75 percent of all meningitis in infants and young children.

C. Select the correct statements about *Streptococcus penumoniae*.

1. It is an encapsulated pathogen.

2. It causes 90 percent of the cases of meiningitis in adults over age 40.

3. Alcoholics are very vulnerable to its infection.

4. It enters the CNS from the digestive tract.

D. Select the correct statement about *E. coli*.

1. It causes less than 5 percent of the bacterial meningitis cases.

2. 90 percent of patients with meningitis and infected with it are cured with the antibiotics ampicillin or gentamicin.

Infections of the Meninges: Viral and Fungal Causes

A. Select the correct statements about aseptic meningitis.

1. It is fairly common.

2. It is caused by one specific virus.

3. Antimicrobial treatment is effective against it.

4. It is a deadly disease.

5. It should be treated with antibiotics until bacterial meningitis is ruled out.

B. Select the correct statements about *Cryptococcus neoformans*.

1. It reduces the leukocyte count per cubic mm.

2. It is a bacterium.

3. It is dimorphic.

4. It is inhaled into the lungs.

Diseases of Neural Tissues: Bacterial Causes

A. Complete each of the following statements about *Clostridium tetani* with the correct term.

1. It is a Gram-_________ rod.

2. It moves by _________ flagella.

3. This bacterium produces ________ under harsh conditions.

4. It grows best at a temperature of __________ degrees Celsius.

5. It grows best at a pH of ________.

6. The wound it infects becomes ________ when the blood supply to it is cut off.

7. It produces a(n) ________, which has a deadly effect on the nervous system.

8. Tetanospasmin affects neurons that stimulate __________.

9. __________ is another name for lockjaw.

10. The treatment for lockjaw is the injection of ________.

B. Complete each of the following statements about *Clostridium botulinum* with the correct term.

1. Its cells are ________-shaped.

2. It produces heat-resistant ________.

3. The botulinum toxin is designated A through ________.

4. Type _________ toxin is the most potent toxin.

5. Its toxin prevents the release of __________ from neurons.

6. It produces ________ paralysis in muscles.

7. Most adults acquire botulism from ________.

8. The process of __________ can destroy its toxin.

Diseases of Neural Tissue: Viral Causes/Prion Causes

A. Select the correct statements about the rabies virus.

1. It contains a single minus strand of RNA.
2. It is usually transmitted by an animal bite.
3. Its earliest phase of rabies is the prodromal phase.
4. The excitation phase is the last phase of rabies.
5. Pasteur's treatment is still used commonly to treat this disease.

B. Select the correct statements about polio and the polio virus.

1. The virus is a DNA virus
2. The virus has seven different serotypes.
3. Cases of polio were reported in the Western Hemisphere in 1996.
4. OPV is the Sabin vaccine.
5. IPV is the Salk vaccine.
6. Polio has been eradicated in the United States.

C. Select the correct statements about the arboviruses.

1. They are a diverse group of RNA viruses.
2. The different types of encephalitis are caused by one arbovirus.
3. All of the viruses are transmitted by mosquitoes.

4. Diagnosis of the encephalitis caused by an arbovirus is difficult.

5. Treatments are available for the diseases caused by the arboviruses.

D. Select the correct statements about the prions and the diseases they cause.

1. The different prion diseases have the same mode of infection.

2. CJD is not a heriditary disease.

3. At least six prion diseases afflict animals.

4. One prion disease is BSE.

Diseases of Neural Tissue: Protozoal Causes

A. Select the correct statements about trypanosome diseases.

1. Two subspecies cause African sleeping sickness.

2. The trypanosomes are transmitted to humans by the bite of the tsetse fly.

3. The trypanosomes damage the brain tissue.

4. Compounds that treat the disease must cross the blood brain barrier.

Multiple Choice: Review

1. The innermost layer of the meninges is the

A. arachnoid mater.
B. cerebral cortex.
C. dura mater.
D. pia mater.

2. Encephalitis is an inflammation of the

A. brain.
B. heart.
C. kidney.
D. skeletal muscles.

3. Select the incorrect statement about *N. meningitidis*.

   A. It avoids the defenses of the human body.
   B. It forms endospores.
   C. It grows well in chocoate agar.
   D. It is Gram-negative.

4. *C. neoformans* enters the body by

   A. breakage in the skin.
   B. contaminated food.
   C. inhalation.
   D. urinary tract infections.

5. Tetanospasmin is produced by a species of

   A. *Bacillus.*
   B. *Clostridium.*
   C. *Escherichia.*
   D. *Streptococcus.*

6. Which strains of *C. botulinum* cause human disease?

   A. A, B, C
   B. A, B, E
   C. C, D, E
   D. D, E, F

7. Select the incorrect statement about the rabies virus.

   A. It belongs to the rhabdovirus family.
   B. It infects many different animals.
   C. It is a DNA virus.
   D. It is bullet-shaped.

8. The Pasteur treatment involves injections into the ________ wall.

   A. abdominal
   B. pericardial
   C. periosteal
   D. pleural

9. The Sabin vaccine is the

   A. IPV.
   B. OPV.

10. African sleeping sickness is caused by the infection of a

    A. bacterium.
    B. fungus.
    C. protozoan.
    D. virus.

Answers

The Nervous System

A. 1. False; Neuroglial cells support neurons and bind them together.
   2. True
   3. False; The dura mater is the outermost meningeal layer.
   4. False; The CSF circulates between the arachnoid and pia mater.
   5. True
   6. True
   7. False; Myelitis is an inflammation of the spinal cord.
   8. True

Infections of the Meninges: Bacterial Causes

A. 2, 3, 4, 6, 8
B. 1, 4
C. 1, 3
D. 1

Infections of the Meninges: Viral and Fungal Causes

A. 1, 5
B. 1, 4

Diseases of Neural Tissues: Bacterial Causes

A. 1. positive 2. peritrichous 3. endospores 4. thirty-seven 5. seven point four 6. anaerobic 7. endotoxin 8. muscle contraction 9. trismus 10. tetanus immune globulin

B. 1. rod 2. endospores 3. G 4. A 5. acetylcholine 6. flaccid 7. food poisoning 8. heating

Chapter 25

Diseases of Neural Tissues: Viral Causes/Prion Causes

A. 1, 2, 3
B. 4, 5, 6
C. 1, 3, 4
D. 3, 4

Diseases of Neural Tissues: Protozoal Causes

A. 1, 2, 3, 4

Multiple Choice: Review

1. D 2. A 3. B 4. C 5. B 6. B 7. C 8. A 9. B 10. C

Chapter 26
Infections of the Body's Surfaces

- The Body's Surfaces
    - Structure and Function of the Skin
    - Structure and Function of the Eye's Surface
    - Defense and Normal Biota: A Brief Review
    - Clinical Syndromes
- Skin Infections
    - Bacterial Causes
        - *Streptotococcus pyrogenes*: Impetigo, Erysipelas
            - Virulence and Pathogenesis
            - Treatment and Prevention
        - *Staphylococcus aureus*: Impetigo, Boils, Abscesses
            - Pathogenesis
            - Treatment
        - *Pseudomonas aeruginosa*: Folliculitis, *Pseudomonas* Infection
        - *Clostridium perfringens*: Gas Gangrene
        - Acne
        - Leprosy (Hansen's Disease)
            - Clinical Syndrome
            - Epidemiology and Treatment
    - Viral Causes
        - Varicella Zoster: Chickenpox, Shingles
            - Clinical Syndromes
            - Prevention and Treatment
        - Fever Blisters, Gingivostomatitis
        - Measles (Rubeola)
            - Epidemiology
            - Subacute Sclerosing Panencephalitis
            - Rubella (German Measles)
            - Smallpox
            - Warts
    - Fungal Causes
        - The Dermatophytes: Ringworm
        - *Candida albicans*: Candidiasis
    - Arthropod Causes
        - Scrabies
        - Pediculosis (Lice)
- Eye Infections
    - Bacterial Causes
        - Inclusion Conjunctivitis, Trachoma
        - Neonatal Gonorrheal Ophthalmia
    - Helminthis Causes
        - Orchocerciasis (River Blindness)

Loa loa (Loaiasis)
Summary

Key Terms

stratum germinativum
keratin
stratum corneum
dermis
sebum
conjunctiva
keratitis
keratoconjunctivitis
impetigo
erysipelas
streptococcal gangrene
leukocidin
exfoliative toxin
scalded skin syndrome
folliculitis
carbuncle
abscess
cellulitis
debridement
cystic fibrosis
gas gangrene
crepitance
isotretinoin
leprosy
spinal ganglia
Reye's syndrome
dermatome
herpetic whitlow
Koplik spots
tinea
candidiasis
trachoma
microfilariae

Study Tips

1. Information is always more meaningful if you can organize it as you study it. For most of the information in this chapter, try organizing it by this table:

Diseases	Microorganism

2. Add to your knowledge of the skin by studying it through additional texts. The free surfaces of the human body are covered by four different kinds of membranes: cutaneous (skin), synovial, serous, and mucous.

3. Are there resources in your biology lab that will add to your study of the concepts of this chapter? Possibilities include studying a model of the skin, a prepared slide of the skin under the microscope, and a model of the eye.

The Body's Surfaces

A. Complete each of the following statements with the correct term.

1. The stratum ________ is the innermost layer of the epidermis of the skin.

2. ________ is a waterproof protein of the skin

3. The ________ is the layer of the skin containing blood vessels.

4. ________ is an enzyme from the skin that kills Gram-positive bacteria.

5. The surface of the eye exposed to the environment is covered with a membrane called the ________.

6. Most of the bacteria living on the skin are not pathogens but are ________.

7. __________ is an infection of the cornea.

Skin Infections

A. The following statements are about skin infections caused by bacteria. Label each statement as true or false. If false, correct it.

1. Impetigo affects the deeper layers of the dermis.

2. Impetigo is caused by two different species of bacteria that form chains and stain Gram-positive.

3. Erysipelas affects only the superficial layers of the skin.

4. Leukocidins destroy red blood cells.

5. Streptolysins destroy white blood cells.

6. *S. pyrogenes* is susceptible to small doses of penicillin.

7. The exfoliate toxin causes the layers of the skin to separate and peel.

8. Cellulitis can lead to life-threatening infections.

9. Penicillinase-producing strains of *S. aureus* produce penicillin.

10. *P. aeruginosa* is an opportunistic pathogen.

11. *P. aeroginosa* causes cystic fibrosis in children.

12. Otitis media is swimmer's ear.

13. *C. perfringens* is an anaerobic, Gram-positive bacterium.

14. Acne is caused by the suppression of an inflammatory response.

15. The tetracyclines are very effective at treating acne.

16. *M. leprae* grows best in the warmer regions of the body.

17. Indeterminate leprosy usually develops three to five years after infection.

18. The rate of leprosy outbreaks is decreasing in the U.S.

B. The following statements are about skin infections caused by viruses. Label each statement as true or false. If false, correct it.

1. VZV infects only humans.

2. Varicella develops along the dermatomes of the body.

3. The Oka strain has been developed to treat smallpox.

4. Acyclovir is used routinely to treat oral HSV infection.

5. The protein projections off the lipid envelop of the rubeola virus allow it to agglutinate red blood cells.

6. Recovery from measles mainly depends on the activity of B lymphocytes.

7. Measles is almost exclusively a disease among children.

8. There is no treatment for SSPE.

9. Rubella is more communicable than measles or chickenpox.

10. The smallpox virus is a small, RNA virus.

C. The following statements are about skin infections caused by fungi. Label each statement as true or false. If false, correct it.

1. Warts is caused by a fungus.

2. The fungi causing ringworm produce an enzyme that destroys keratin.

3. Griseofulvin specifically targets the dermis of the skin.

4. *Candida* species are opportunistic pathogens.

5. *Candida* species thrive in a dry environment.

6. Scabies is confined mainly to the United States.

7. Scabies is a harmless infection.

8. Lice are usually transferred by direct bodily contact.

Eye Infections

A. Short Answer

1. How is the exposed surface of the eye protected?

2. Which bacterium is the most usual cause of conjunctivitis?

3. When trachoma develops in the eye, where do the bacteria normally multiply?

4. What is the treatment for a scarred cornea?

5. How is blindness from *N. gonorrhoeae* prevented in a newborn?

6. Which viruses commonly infect the conjunctiva of the eye?

7. What is herpetic keratitis?

8. How is river blindness transmitted?

9. What is suramin?

10. Where does loaiasis occur througout the world?

Discussion Questions

1. How is the skin an important first line of defense at warding off invading microorganisms. Are there other first lines of defense?

2. How do the previous lessons in microbiology support the concepts of this chapter?

Multiple Choice: Review

1. Select the incorrect association.

   A. conjunctiva/epidermis and connective tissue
   B. keratitis/cornea
   C. stratum corneum/outer layer
   D. stratum germinativum/dermis

2. Select the incorrect statement about *S. pyrogenes*.

   A. It can only infect certain organs of the body.
   B. It is Gram-positive.
   C. It is susceptible to small doses of penicillin.
   D. It infects the skin.

3. Burns are most commonly infected by a species of

   A. *Bacillus*.
   B. *Clostridium*.
   C. *Pseudomonas*.
   D. *Streptococcus*.

4. Select the incorrect statement about *Clostridium perfringens*.

   A. It is abundant in the soil.
   B. It is anaerobic.
   C. It is destroyed by necrotic tissue.
   D. It is Gram-positive.

5. Leprosy is caused by a species of

   A. *Escherichia.*
   B. *Mycobacterium.*
   C. *Staphylococcus.*
   D. *Streptococcus.*

6. Chickenpox is caused by a

   A. bacterium.
   B. fungus.
   C. protozoan.
   D. virus.

7. Dermatomes correspond to the distribution of __________ in the body.

   A. blood vessels
   B. brain lobes
   C. fat storage areas
   D. sensory nerves

8. Cell-mediated immunity is carried out by sensitized

   A. T lymphocytes.
   B. B lymphocytes.

9. Which virus is the most powerful teratogen?

   A. measles
   B. mumps
   C. rubella
   D. smallpox

10. Ringworm is caused by a

   A. bacterium.
   B. fungus.
   C. helminth.
   D. virus.

Answers

The Body's Surfaces

A. 1. germinativum

2. keratin
3. dermis
4. lysozyme
5. conjunctiva
6. commensals
7. keratitis

Skin Infections

A.
1. False; It is a superficial infection.
2. False; It is caused by *Staphylococcus aureus* and *Staphylococcus pyrogenes*. These bacteria occur in clusters (staphylococcus).
3. False; It affects the dermis.
4. False; They destroy white blood cells.
5. False; They destroy red blood cells.
6. True
7. True
8. True
9. False; They produce an enzyme against penicillin.
10. True
11. False; It infects children who have developed cystic fibrosis.
12. False; Otitis externa is swimmer's ear.
13. True
14. False; It is caused by an inflammatory response.
15. True
16. False; It grows in the cooler regions of the body.
17. True
18. False; It is rare but is increasing in frequency.

B.
1. True
2. True
3. False; It is a vaccine to combat varicella zoster.
4. False; It is used to treat genital herpes, but not used for this disease.
5. True
6. False; It is fought by cell-mediated immunity through T lymphocytes.
7. True
8. True
9. False; It is not as communicable as these diseases.
10. False; It is a large, double-stranded DNA virus.

C.
1. False; Warts is caused by a virus.
2. True
3. False; It concentrates on the stratum corneum.
4. False; They are commensal fungi.
5. False; They thrive in moist environments.

6. False; It is common worldwide.
7. True
8. True

Eye Infections

A. 1. It is protected by the tear fluid.
2. The bacterium is *Pseudomonas aeruginosa*.
3. They multiply on the conjunctiva.
4. The answer is a corneal transplant.
5. It is treated by the administration of erythromycin.
6. The viral groups are the adenoviruses and enteroviruses.
7. It is an ulceration of the cornea.
8. It is transmitted by black flies and buffalo gnats of *Simulium*.
9. It is an anti-helminth drug used to treat river blindness.
10. It occurs only in Africa.

Discussion Questions

1. The skin presents a physical barrier and the sebaceous glands secrete fatty acids that also establish a protective layer. Other first lines of defense include the acidity of the stomach and the mucous membrane-ciliary action of the respiratory tract.

2. Tie-ins include bacterial morphology and metabolism along with bacterial nutrition and genetics. The genetics of a microorganism establishes its biochemical makeup and potential to infect. The human body offers a source of nutrition as the microorganism infects.

Multiple Choice: Review

1. D 2. A 3. C 4. C 5. B 6. D 7. D 8. A 9. C 10. B

Chapter 27
Systemic Infections

- The Cardiovascular and Lymphatic Systems
  - Structure and Function
    - The Cardiovascular System
    - The Lymphatic System
  - Defenses and Normal Biota: A Brief Review
  - Clinical Syndromes
- Infections of the Heart
  - Endocarditis
    - Bacterial Endocarditis
    - Acute and Subacute Endocarditis
  - Myocarditis
    - Viral Myocarditis
    - American Trypanosomiasis (Chagas' Disease)
  - Pericarditis
- Systemic Infections
  - Bacterial Infections
    - Plague
    - Tularemia
    - Brucellosis
    - Lyme Disease
    - Relapsing Fever
    - Anthrax
    - Cat Scratch Disease
    - Rocky Mountain Spotted Fever
    - Typhus
      - Epidemic (Louseborne) Typhus
      - Murine (Fleaborne) Typhus
  - Viral Infections
    - Yellow Fever
    - Dengue Fever
    - Infectious Mononucleosis
      - Pathogenesis
      - EBV-Associated Cancers
    - Acquired Immunodeficiency Syndrome (AID)
      - Untreated HIV Infections
      - Treated HIV Infections
      - AIDS and Other Diseases
      - Transmission
      - Epidemiology
      - Prevention
    - Ebola Hemorrhagic Fever
      - The Disease

Key Terms

atrium
ventricle
valve
pulmonary circulation
systemic circulation
aorta
artery
capillary
vein
endocardium
myocardium
pericardium
septicemia
bubonic plague
transovarially
ulceroglandular
cutaneous anthrax
anthrax
infectious mononucleosis
Burkitt's lymphoma
viral load
blackwater fever
miracidia
cercaria
elephantiasis

Chapter 27

Study Tips

1. Much of the information in this chapter relates infections of the cardiovascular and lymphatic systems to causative microorganisms. Organize this information by composing the following table:

Disease	Microorganism

2. Consult additional sources to study the anatomy and physiology of the cardiovascular and lymphatic systems in more depth. This will promote your understanding of the infections of these systems.

3. Is a human torso model available in your lab? Is a model of the heart also available? Use these as aids to study the human anatomy described in this chapter.

4. Continue to answer the questions at the end of the chapter in the text. After answering them, ask your instructor to provide a key for the answers.

The Cardiovascular and Lymphatic Systems

A. Complete each of the following statements with the correct term or terms..

1. A heart is dissected in a biology lab. One of the chambers examined has a thick wall. The largest vessel attached to the heart attaches to this chamber. The chamber is the ________ ________.

2. The action of the valves prevents the __________ of blood through the heart.

3. Blood returning to the heart from the lungs has a comparatively higher concentration of the gas _________. It has a comparatively lover concentration of the gas ________.

4. A large blood vessel with a comparatively thin wall is discovered in the dissection of a mammal. This blood vessel is a ________.

5. The ___________ are the microscopic blood vessels of the circulatory system.

6. In one example of a disease, fluid collects around the heart. This fluid is trapped by the ________ membrane.

Infections of the Heart

A. Complete each of the following statements with the correct terms or terms.

1. A bacterium adheres to the inside free surface of the heart. It is attaching to the _________ of the heart.

2. ________ are bulky masses of bacteria and clots.

3. Among intravenous drug users with bacterial endocarditis, the bacterium ________ _________ causes over one-half of the cases of endocarditis.

4. In the U.S. most of the cases of endocarditis are caused by ________ microorganisms.

5. Another name for trypanosomiasis is _________ disease.

6. Cells vulnerable to trypanosomiasis include brain cells, ________ cells, and ________ cells.

7. _________ are the most common cause of viral endocarditis.

8. _________ is the surgical removal of the pericardium.

Systemic Infections

A. The following statements are about bacterial infections. Label each statement as true or false. If false, correct it.

1. Humans are the main reservoir for *Yersinia pestis*.

2. To cause the plague, *Yersinia pestis* enters the human lymphatic system.

3. The bubonic plague is more virulent than the pneumonic plague.

4. Tularemia only occasionally infects humans.

5. Bacteria causing tularemia are killed in the human body by phagocytosis.

6. Species of *Brucella* are Gram-negative coccobacilli.

7. Species of *Brucella* are able to survive as intracellular parasites in phagocytes.

8. Arthritis develops is a person within one month after being bitten by the tick that spreads Lyme disease.

9. Lyme disease can be eradicated from a geographical area if deer are removed from that area.

10. Lyme disease is treated intravenously by use of tetracycline.

11. Epidemic and endemic are two phases of relapsing fever.

12. The causative microorganism of anthrax is a species of *Pseudomonas*.

13. There are respiratory and gastrointestinal forms of anthrax.

14. *Bartonella henselae* is probably a normal oral biota of cats and dogs.

15. *Rickettsia ricketsii* normally becomes concentrated in the alveoli of the lungs when it infects the body.

16. Early detection of RMSF is relatively easy.

17. Early diagnosis of typhus is difficult.

18. Only humans and lice are necessary to maintain typhus.

19. Murine typhus is endemic to Western Europe.

20. Typhus is caused by a virus.

B. The following statements are about viral infections. Label each of the following statements as true or false. If false, correct it.

1. Yellow fever is caused by *A. aegypti*.

2. A vaccine exists to prevent dengue fever.

3. Infectious mononucleosis occurs frequently in young children.

4. EBV typically infects the neurons of the body.

5. EBV establishes a latent infection in T lymphocytes.

6. The EBV can be oncogenic.

7. HIV infection usually begins in red blood cells.

8. From viral load a person has AIDS when the CD+4 T cell count drops to 600.

9. HAART contains reverse transcriptase and protease inhibitors.

10. A secondary infection is usually the immediate cause of a death from AIDS.

11. AIDS can be transmitted by the bite of an insect vector.

12. In Africa the death rate from AIDS is nearly equal among males and females.

13. The HIV has a remarkable ability to mutate.

14. The sole reservoir of the Ebola virus is the human population.

15. The Ebola virus belongs to the Filoviridae.

C. The following statements are about protozoan infections. Label each of the following statements as true or false. If false, correct it.

1. Most people who suffer from malaria are children.

2. After introduction by a mosquito bite, the *Plasmodium* is converted from a merozoite to a sporozoite in the liver.

3. The gametocytes of *Plasmodium* develop from merozoites in the human blood.

4. The release of malarial antibodies causes the periodic chills in the human.

5. The protozoan causing malaria is introduced into the human body through the bite of the *Culex* mosquito.

6. Today malaria is confined to tropical and subtropical areas.

7. Natural immunity to malaria is strong.

8. Tachyzoites multiply rapidly in mature red blood cells.

9. *Toxoplasma* cysts are common in meat.

10. A tick vector transmits the parasite for babesiosis from one host to another.

D. The following statements are about helminth infections. Label each of the following statements as true or false. If false, correct it.

1. Shistosome eggs hatch directly into cercariae larvae.

2. Cercariae larvae directly infect people to cause shitosomiasis.

3. Shistosomes are prevalent in Asia, Africa, and the Middle East.

4. Filariasis is introduced into the human body through the bite of a tick.

5. In the development of elephantiasis, the circulation of the systemic blood flow is blocked.

Chapter 27

Discussion Questions

1. Capillaries are exchange vessels of the body. How do you think this exchange function is lost in the development of edema?

2. How do the principles of human anatomy and physiology support the understanding of the concepts in this chapter on systemic infection?

Multiple Choice: Review

1. Pulmonary veins

   A. return blood to the heart from regions throughout the body.
   B. return blood to the heart from the lungs.
   C. transport blood from the heart to various body regions.
   D. transport blood from the heart to the lungs.

2. Endocarditis in humans is mainly

   A. bacterial and affects the inner layer of the heart.
   B. bacterial and affects the outer layer of the heart.
   C. viral and affects the inner layer of the heart.
   D. viral and affects the outer layer of the heart.

3. Trypanosomiasis is caused in humans by a

   A. bacterium.
   B. fungus.
   C. protozoan.
   D. virus.

4. *Yersinia pestis* is a

   A. Gram-negative coccibacillus.
   B. Gram-negative spirochete.
   C. Gram-positive coccibacillus.
   D. Gram-positive spirochete.

5. Occuloglandular disease infects mainly the

   A. brain.
   B. eye.
   C. heart.
   D. kidney.

6. Select the incorrect statement about *B. burgdoferi*.

   A. It causes Lyme disease.
   B. It has mutated frequently.
   C. It is a spirochete.
   D. It is easy to cultivate in the laboratory.

7. Anthrax is caused by a species of

   A. *Aerobacter*.
   B. *Bacillus*.
   C. *Pseudomonas*.
   D. *Staphylococcus*.

8. Select the incorrect statement about *Rickettsia ricketsii*.

   A. It causes RMSF.
   B. It is Gram-negative.
   C. It is rod-shaped.
   D. It mainly infects neurons.

9. Yellow fever is caused by a

   A. bacterium.
   B. fungus.
   C. protozoan.
   D. virus.

10. The nasopharyngeal carcinoma caused by EBV is common in

   A. Africa.
   B. China.
   C. India.
   D. South America.

Answers

The Cardiovascular and Lymphatic Systems

A. 1. left ventricle
   2. backflow
   3. more oxygen, less carbon dioxide
   4. vein
   5. capillaries
   6. pericardial

Chapter 27

Infections of the Heart

A. 1. endocardium
2. vegetations
3. *Staphylococcus aureus*
4. viral
5. Chagas'
6. liver, myocadial
7. enteroviruses
8. pericardectomy

Systemic Infections

A. 1. False; The reservoir is different species of rodents.
2. True
3. False; Pneumonic plague is more virulent.
4. True
5. False; They are phagocytized by not killed.
6. True
7. True
8. False; Arthritis develops six months after the tick bite.
9. True
10. False; Drugs used are doxycycline, amoxicillin, and erythromycin.
11. True
12. False; The causative microorganism is *B. anthracis*.
13. True
14. True
15. False; It is found along the inner lining of blood vessels.
16. False; Early detection is difficult.
17. True
18. True
19. False; It is endemic in the U.S.
20. False; It is caused by a bacterium.

B. 1. True
2. False; A vaccine is not available.
3. False; Infection usually does not occur until the age span of 15 to 25.
4. False; If infects epithelial cells near the salivary glands.
5. False; It produces the latent infection in B lymphocytes.
6. True
7. False; Surface (epithelial) cells become infected first.
8. False; It occurs when this count drops to 200.
9. True
10. True
11. False; It is not one of the three modes of transmission.

12. True
13. True
14. False; It is an unknown virus in the tropical rain forest.
15. True

C. 1. True
2. False; A sporozoite is converted to a merozoite.
3. True
4. False; The release of tumor necrotizing factor causes the chills.
5. False; It is acquired from the bite of the *Anopheles* mosquito.
6. True
7. False; It is a weak response.
8. False; They invade all body cells except mature red blood cells.
9. True
10. True

D. 1. False; They hatch into miracidia larvae.
2. True
3. True
4. False; It is introduced by the bite of an insect.
5. False; It blocks lymphatic circulation.

Discussion Questions

1. The volume of tissue fluid returned to the capillary does not equal the volume of fluid leaving the capillary to supply tissue cells. Therefore, this fluid collects in the tissue spaces. The return of tissue fluid is inhibited due to the loss of plasma proteins from the capillaries into the tissue spaces. Therefore, the fluid is not returned sufficiently to the capillary by osmosis.

2. Start with examples of how nutrients are absorbed across the mucosal wall of the GI tract. Also study the rate of clearance of drugs from the blood by renal function.

Multiple Choice: Review

1. B 2. C 3. C 4. A 5. B 6. B 7. B 8. D 9. D 10. B

Chapter 28
Microorganisms and the Environment

Life and the Evolution of Our Environment
Microorganisms in the Biosphere
    Soil
        Mineralization
        Microorganisms in the Soil
            Bacteria
            Fungi
            Other Microorganisms
        Symbiosis
            Mycorrhizae
            The Rhizosphere
        Water
            Nutrients
            Pathogens
        Air
The Cycles of Matter
    The Nitrogen Cycle
        Nitrogen Fixation
        Nitrification
        Denitification
    The Carbon Cycle
    The Phosphorus Cycle
    The Sulfur Cycle
        Reduction of Sulfate
        Oxidation of Hydrogen Sulfide
        Products and Reactions
Treatment of Waste Water
    Sewage Treatment Plants
    Septic Tanks and Oxidation Ponds
Treatment of Drinking Water
    Processing Methods
    Testing Methods
Escape of the Carbon Cycle
Summary

Key Terms

biogeochemical transformations
mineralization
soil fertility
symbiosis

mutualism
mycorrhizae
rhizosphere
rhizosphere effect
R:S ratio
eutrophic
aerosol
nitrogen fixation
nitrogenase
root hair
nitrogen fixation
ammonification
nitrification
denitrification
desulfurylation
biochemical oxygen demand
trickling filter
activated sludge
methanogen
septic tank
leach field
sewage farming
coliform bacterium
MNP test
recalcitrant organic compounds

Study Tips

1. Answer the different kinds of questions at the end of the chapter. Answer them in your own words after studying the chapter. After answering them, then check the chapter content and compare it with your answers.

2. Review the chapters on basic chemistry (Chapter 2) and prokaryotic species (Chapter 11). How does the information in these chapters help you understand the concepts in Chapter 28?

3. Study the various biogeochemical cycles in this chapter. After studying them, try to outline each cycle in your own words and by your own patterns.

4. Is there a local plant that treats waste water? Visit it and write a short report on your findings. Talk to your instructor about making this an extra credit project.

Chapter 28

Life and the Evolution of Our Environment

A. Label each of the following statements as true or false. If false, correct it.

1. Sulfur is a usable, major biological element.
2. The Earth was formed about 3.5 billion years ago.
3. The early atmosphere of the Earth lacked oxygen.
4. The oldest fossils known resemble protozoa.
5. Today oxygen comprises about 30 percent of the atmosphere of the Earth.

Microorganisms in the Biosphere

A. Complete each of the following with the correct term.

1. The _________ is the region of the Earth that supports life.
2. The process of ___________ converts organic material to inorganic material.
3. Soil fertility depends on an adequate supply of the elements ________, _________, and ________.
4. Most soils contain __________ microorganisms, as the soil can become hot during the day.
5. Actinomycetes are Gram-________ bacteria.
6. The bacterium ________ __________ is an opportunist that causes infections in burn victims.
7. __________ is a symbiotic relationship that benefits both organisms involved.
8. ___________ is a symbiotic association between fungi and the roots of plants.
9. The ________ ratio is the ratio of the concentration of microorganisms in the rhizosphere to their concentration in the adjacent soil.
10. *Pseudomonas* has Gram-__________ species.
11. About __(number)__ of the water of the Earth is seawater.
12. Most photosynthesis is conducted by ___________ in the water.
13. __________ waters are enriched with nutrients.

14. Microorganisms do not grow in the ________.

15. __________ are tiny particles of liquid.

The Cycles of Matter

A. Match each of the following descriptions to the correct biogeochemical cycle.

1.	ammonia is converted	A.	carbon cycle
2.	its gas is a small part of the air	B.	nitrogen cycle
3.	gaseous intermediate is lacking	C.	phosphorus
4.	involves tenth most abundant element	D.	sulfur

B. Complete each of the following statements with the correct term.

1. Nitrogen makes up about ________ percent of the atmosphere of the Earth.
2. The process of ________ ________ converts nitrogen from the atmosphere into a usable form.
3. ___________ is an enzyme catalyzing nitrogen fixation.
4. The _________ process industrially fixes nitrogen.
5. Nitrification converts ammonia to the ________ ion.
6. ___________ is the main form of nitrogen used by plants.
7. ________ is the conversion of nitrate into nitrogen gas.
8. ___________ destroys the ozone layer of the atmosphere.
9. Carbon dioxide makes up about __(number)__ percent of the atmosphere of the Earth.
10. Carbon is present in the atmosphere mainly as __________.
11. ________ are organisms that generally fix carbon dioxide from the atmosphere by photosynthesis.
12. __________ is the metabolic process that returns carbon dioxide to the atmosphere.
13. Carbon is dissolved in waterways as the ________ion.

14. Phosphorous is present in the ocean as the ________ion.

15. Phosphorus lacks a ________ intermediate as it cycles

16. In short supply, phosphorus is a ________ nutrient in most soils.

17. By the sulfur cycle sulfur passes through various ________ states.

18. Most microorganisms derive their sulfur from the ________ ion.

19. By ________ organic sulfur compounds are converted to hydrogen sulfide.

20. Sulfur-oxidizing bacteria oxidize sulfide to ________.

Treatment of Waste Water

A. Label each of the following statements as true or false. If false, correct it.

1. Sewage is municipal waste water.

2. Sewage treatment plants add oxygen to sewage at a slower rate than microbes remove it by metabolism.

3. Primary sewage treatment is biological rather than mechanical.

4. Secondary sewage treatment is mechanical rather than biological.

5. The effluent is the liquid entering the sewage treatment plant.

6. Digested sludge is the solid portion of primary sludge remaining after primary treatment.

7. Most tertiary treatments of waste are chemical rather than biological.

8. A septic tank is a small anaerobic digester.

9. Waste water is classified by its BOD.

10. Methanogens use methane.

Treatment of Drinking Water

A. Complete each of the following statements with the correct term.

1. To clarify water is to remove ____________ from it.

2. Ozone or ________ is added to water supplies to kill microorganisms.

3. ________ bacteria are used as indicators for the quality of drinking water.

4. The first test of the MPN test is the ________ test.

5. By the MF procedure microbes are cultured on a(n) ________ plate.

Escape From the Carbon Cycle

1. What are recalcitrant compounds?

Discussion Questions

1. Ecology is a branch of biology that explains the relatedness of organisms. How are principles of this subject demonstrated in this chapter?

2. What are the ecological problems when materials are not biodegradable?

Multiple Choice: Review

1. Select the correct description for humus.

   A. inorganic, brown or black
   B. inorganic, red or yellow
   C. organic, brown or black
   D. organic, red or yellow

2. Select the correct description of the fungi.

   A. aerobic, decomposers
   B. aerobic, producers
   C. anaerobic, decomposers
   D. anaerobic, producers

3. Which kind of symbiotic relationship benefits both organisms?

   A. commensalism
   B. mutualism
   C. opportunism
   D. parasitism

4. Water covers about ________ percent of the surface of the Earth.

   A. 50
   B. 60
   C. 70
   D. 80

5. The conversion of nitrates into nitrogen gas is

   A. ammonification.
   B. denitrification.
   C. nitrification.
   D. nitrogen fixation.

6. Carbon dioxide is returned to the atmosphere by

   A. photosynthesis.
   B. respiration.

7. The main inorganic supply of phosphorus is in

   A. rocks.
   B. the atmosphere.
   C. the ocean.
   D. the streams.

8. Desulfurylation converts organic sulfur compounds to

   A. hydrogen sulfide.
   B. sulfate.
   C. sulfuric acid.
   D. sulfuric acid.

9. The tertiary treatment of waste water is usually

   A. biological.
   B. chemical.
   C. enzymatic.
   D. mechanical.

10. The second step of the MPN procedure is the ________ test.

   A. coliform
   B. completed
   C. confirmed
   D. presumptive

Answers

Life and the Evolution of Our Environment

A. 1. True
   2. False; The Earth was formed about 4.6 billion years ago.
   3. True
   4. False: The oldest known fossils resemble cyanobacteria.
   5. False; Today oxygen comprises about 21 percent of the atmosphere of the Earth.

Microorganisms in the Biosphere

A. 1. biosphere 2. mineralization 3. sodium, potassium, phosphorus 4. thermophilic 5. positive 6. *Pseudomonas aeroginosa* 7. mutualism 8. mycorrhizae 9. R:S 10. negative 11. ninety-nine 12. phytoplankton 13. eutrophic 14. air 15. aerosols

The Cycles of Matter

A. 1. B 2. A 3. C 4. D

B. 1. eighty 2. nitrogen fixation 3. nitrogenase 4. Haber 5. nitrate 6. nitrate 7. denitrification 8. denitrification 9. 0. 03 10. carbon dioxide 11. autotrophs 12. respiration 13. carbonate 14. phosphate 15. gaseous 16. limiting 17. oxidation 18. sulfate 19. desulfurylation 20. sulfate

Treatment of Waste Water

A. 1. True
   2. False; They do the opposite of this.
   3. False; It is mechanical rather than biological.
   4. False; It is biological rather than mechanical.
   5. False; It is liquid leaving the treatment plant.
   6. True
   7. True
   8. True
   9. True
   10. False; They are bacteria that make methane.

Treatment of Drinking Water

A. 1. pathogenic microorganisms 2. chlorine 3. coliform 4. presumptive 5. EMB

Chapter 28

Escape from the Carbon Cycle

A. 1. There compounds are not biodegradable and their elements cannot be released for recycling in the environment.

Discussion Questions

1. Start with an example of phosphorus as a limiting factor in most waterways. How is life changed here if phosphorus becomes more abundant?

2. Materials that are not biodegradable have elements that will not cycle in the environment. This keeps them tied up and they cannot be reused.

Multiple Choice: Review

1. C 2. C 3. B 4. C 5. B 6. B 7. A 8. A 9. B 10. C

Chapter 29
Microbial Biotechnology

- Traditional Uses of Microorganisms
    - Lactic Acid Bacteria
        - Plant Foods
        - Cheese and Other Dairy Products
    - Yeasts
        - Wine
        - Other Alcoholic Beverages
        - Vinegar
        - Bread
    - Mixed Cultures
- Microbes and Insecticides
- Microbes as Chemical Factories
    - Anaerobic Fermentations
        - Ethanol
        - Acetone and Butanol
    - Aerobic Processes
        - Antibiotics
        - Amino Acids
        - Enzymes
        - Chemical Reactions
- Using Genetically Engineered Microbes
    - Medical Uses
        - Hormones
            - Human Growth Hormone
            - Insulin
        - Other Proteins
    - Industrial Uses
    - Agricultural Uses
        - Bt Toxin
        - Ice-Minus Bacteria
        - Better Silage Makers
- Summary

Key Terms

silage
curdling
ripening
whey
brandy
saccharified

blackstrap molasses
sparger
impeller
penicillinase
intron
reverse transcriptase

Study Tips

1. This chapter shows the application of many concepts that you have learned throughout your study of microbiology. Can you identify these concepts? Review the text on topics such as bacterial metabolism and bacterial genetics.

2. Visit a local brewery or bakery to learn more about the application of microbiology in commercial industry.

3. Does your college have a biologist conducting research that is relevant to genetic engineering? Interview this person to learn more about this topic firsthand.

Traditional Uses of Microorganisms

A. Complete each of the following statements with the correct term or terms.

1. The production of acid by lactic acid bacteria __________ food, as their acid kills or inhibits other microorganisms.

2. The term "sauerkraut" means ________.

3. Lactic acid bacteria of *Lactobacillus* ________ ferment sugar in plant material.

4. Cheese is made by the steps of ________ and ________.

5. Cottage cheese is made from the action of the enzyme _______.

6. Hard cheese is softened by the chemical digestion or ___________ of the proteins in the cheese.

7. Yeasts of the scientific name ________ _________ are used to ferment sugar.

8. ___________ is the process of winemaking.

9. ___________ is the process of grape growing.

10. Either _________ or ____________ wine can be made from red grapes.

11. The metabolic products of the bacterium __________ ______________ give sherry its distinct taste.

12. *Botrytis cinerea* is a ___________ that produces certain sweet wines.

13. By secondary fermentation, the bacterium *Oenococcus oeni* converts _________ acid to lactic acid for winemaking.

14. The enzyme amylase produces a disaccharide from starch. This disaccharide, consisting of bonded glucose molecules, is ________.

15. Beer is flavored from the flowers of ________ ________.

16. Vinegar is a solution of _______ ________.

17. Vinegar is produced by the action of the bacterium ________.

18. ________ __________ is the gas produced by the yeast *S. cerevisiae* to leaven bread.

Microbes as Insecticides

A. Complete each of the following statements with the correct term.

1. Bt is produced by the bacterium ________ _______.

2. Bp is produced by the bacterium ________ ________.

3. Spores of the protozoan ________ _________ are sold commercially as a bait to combat grasshoppers.

Microbes as Chemical Factories

A. Label each of the following statements as true or false. If false, correct it.

1. Fermentations produce large amounts of ATP.

2. The most common substrates for industrial fermentations are blackstrap molasses.

3. Methanol is the alcohol found in alcoholic beverages.

4. Gasohol is 50 percent gasoline and 50 percent alcohol.

5. A species of the genus *Clostridium* has been used commercially to make the solvents acetone and butanol.

6. Most modern industrial microbiological processes are anaerobic.

7. The antibiotic industry began with the work of Alexander Fleming.

8. Penicillin is difficult to synthesize chemically.

9. Most broad-spectrum antibiotics are synthesized commonly in industry.

10. Lysine is added to bread in the United States.

11. Corticosteroids produce an inflammatory response in the human body.

12. Microorganisms are a rich source of enzymes that have many commercial uses.

Using Genetically Engineered Microbes

A. Label each of the following statements as true or false. If false, correct it.

1. hCG is secreted from the posterior pituitary gland.

2. Prokaryotes lack enzymes to eliminate introns.

3. cDNA is produced from reverse transcription.

4. As a technique of genetic engineering, transformation is the introduction of native or recombinant DNA into a host cell.

5. The action of the hormone insulin is to lower glucose in the cell.

6. Human insulin can now be produced from *E. coli*.

7. tPA binds to blood clots and dissolves them.

8. tPA is produced from microbial cells by genetic engineering.

9. Human DNAase is used to treat cystic fibrosis.

10. Vitamin C is produced commercially from plants by harvesting it from the fruits of plants.

11. *P. fluorescens* is a species of bacterium that inhabits the rhizosphere.

12. The presence of *P. syringae* makes frost damage less likely on plants.

Discussion Questions

1. In additions to the applications described in this chapter, what other problems in our society could be solved through the products of genetic engineering?

2. Often scientific investigations take unexpected turns with unexpected discoveries. How was this true in the discovery of antibiotics?

Multiple Choice: Review

1. Select the correct statement about lactic acid bacteria.

   A. They are harmful to humans and decompose food.
   B. They are harmful to humans and preserve food.
   C. They are harmless to humans and decompose food.
   D. They are harmless to humans and preserve food.

2. Hard cheese is made softer by the

   A. formation of proteins through dehydration synthesis.
   B. formation of proteins through hydrolysis.
   C. digestion of proteins through dehydration synthesis.
   D. digestion of proteins through hydrolysis.

3. *Saccharomyces cerevisiae* is a

   A. bacterium that ferments sugars.
   B. bacterium that makes sugars.
   C. yeast that ferments sugars.
   D. yeast that makes sugars.

4. Select the incorrect association.

   A. *Acetobacter*/aerobic bacterium
   B. amylase/hydrolyzes starch
   C. saccharified/produces glucose
   D. vinegar/acetic acid

5. Species of the bacterial genus __________ are insect pathogens.

   A. *Aerobacter*
   B. *Bacillus*
   C. *Escherichia*
   D. *Pseudomonas*

6. Fermentations produce _________ amounts of ATP.

   A. large
   B. small

7. Gasohol is ___1__ percent gasoline and __2__ percent alcohol.

   A. 1 - ninety, 2 - ten
   B. 1 - seventy, 2 - thirty
   C. 1 - fifty, 2 - fifty
   D. 1 - twenty, 2 - eighty

8. In anaerobic processes in industry a rotating shaft is a

   A. fermenter.
   B. impeller.
   C. silager.
   D. sparger.

9. Penicillase is

   A. alpha-lactamase that destroys penicillin.
   B. alpha-lactamase that makes penicillin.
   C. beta-lactamase that destroys penicillin.
   D. beta-lactamase that makes penicillin.

10. A deficiency of insulin in the body will tend to __________ the concentration in the blood.

   A. decrease
   B. increase

Answers

Traditional Uses of Microorganisms

A. 1. preserves
   2. acid cabbage
   3. *plantarum*
   4. curdling, ripening
   5. rennin
   6. hydrolysis
   7. *Saccharomyces cerevisiae*
   8. enology
   9. viticulture
   10. red, white

11. *T. delbrueckii*
12. fungus
13. malic
14. maltose
15. *Humulus lupulus*
16. acetic acid
17. *Acetobacter*
18. carbon dioxide

Microbes as Insecticides

A. 1. *B. thuringiensis*
2. *B. popilliae*
3. *Nosema locustae*

Microbes as Chemical Factories

A. 1. False; Fermentations produce small amounts of ATP compared to aerobic respiration.
2. True
3. False; Ethanol is the alcohol of alcoholic beverages.
4. False; Gasohol is 90% gasoline and 10% alcohol.
5. True
6. False; Most of these processes are aerobic.
7. True
8. True
9. False; They are too complex for chemical synthesis.
10. False; It is not added, as enough is available from the consumption of other foods.
11. False; They suppress inflammation.
12. True

Using Genetically Engineered Microbes

A. 1. False; It is secreted from the anterior pituitary gland.
2. True
3. True
4. False; This process is called transfection.
5. False; It lowers the glucose concentration in the blood.
6. True
7. True
8. False; It is produced from CHO cells.
9. True
10. False; It is produced from chemical steps and microbial fermentation.
11. True

12. False; Its presence makes frost damage more likely.

Discussion Questions

1. Consider the production of erythropoietin as a recent breakthrough to treat anemia. Can you think of other human diseases that could be cured by genetic engineering?

2. Review the research conducted by Fleming. How did chance favor the prepared mind in this instance?

Multiple Choice: Review

1. D 2. D 3. C 4. A 5. B 6. B 7. A 8. B 9. C 10. B